The Carson Factor

ALSO BY WILLIAM ASHWORTH:

Hells Canyon
The Wallowas

The Carson Factor

William Ashworth

F37119

HAWTHORN BOOKS, INC.
Publishers/NEW YORK
A Howard & Wyndham Company

THE CARSON FACTOR

Library of Congress Catalog Card Number: 78-73333

ISBN: 0-8015-3112-8

1 2 3 4 5 6 7 8 9 10

Preface

WHAT IS THE "balance of nature"? Through what forces does it operate, and what objects does it affect? Does it exist at all? If not, why not? If so, what is the nature of "nature"?

All these questions should have been answered long ago, and it is a measure of our ignorance as a species that they have not. We have a firm enough grasp of celestial mechanics to put a man on the moon and a camera on Mars; we know enough chemistry to create cloth out of crude oil, paint out of peanuts, and fertilizers out of thin air; we have split the atom and harnessed the forces that bind its nucleus together to cook our meals, run our television sets, and drive our electric toothbrushes; and we have learned how to manipulate electrons so cleverly that computing equipment that once required the space of two city lots to set up can now be fit quite nicely into a box the size of a standard portable typewriter. But of the forces that control the population densities of living things—those forces that prevent the offspring of a single pair of aphids on your rosebush from multiplying within a few short months to the point where the bulk of them equals the size of your house, as they are perfectly capable, mathematically, of doing—our knowledge has not advanced much beyond the Old Stone Age.

Or perhaps that is putting it too strongly. Perhaps the knowledge is there, and it is only our use of it that lags so abysmally. We have, after all, a pair of perfectly good sciences—ecology and population biology—with the avowed purpose of dealing with questions of natural organization and balance, and these sciences are quite capable of holding their own, both in terms of raw data and in terms of carefully conceived and tested theories based on that data, with the more familiar disciplines such as chemistry and physics. We have the writings of people such as Paul Erlich and Rachel Carson, who have expressed the concepts of ecology and population biology in ways that allow anyone who can read to grasp them, and we have the results of hundreds of experiments and thousands of observations for writers such as these to interpret. We even have a certain amount of popular awareness of all this: Carson's and Erlich's books have been best sellers. Earth Day was an unquestionable success, and the environmental movement, which has made the word *ecology* a kind of battle cry, has become a powerful political force and a potent vehicle for social change. What we do *not* have as yet, on any significant level—and what we desperately need— is a commitment to make practical use of this knowledge in the same manner as we make practical use of chemistry and physics.

We need a radical revision of engineering practice: We get Environmental Impact Statements. We need a fundamentally new definition of progress, and what we get is the same old tired wheeze, progress defined as—in the apparently serious words of the 1978 Annual Report of the El Paso Electric Company—"the art of converting raw land into a new business, subdivision or a new transmission line." It's almost funny. People who wouldn't think of taking away a portion of a building without checking leverage, load factors, sheer strength, compression strength, and a whole host of related data that can tell them whether or not the building will fall down, will casually change the species composition of an ecosystem without a care in the world, paying no attention to food webs, trophic levels, biological magnification, or any other information—just as reliable as structural mechanics, and nearly as easily obtained—that will tell them whether or not the ecosystem will collapse and what the effects will be if it does. Engineers who eagerly sort through the latest research in electronics or atomic physics or molecular chemistry for information that will help them design better radios and more durable roadways and more efficient refrigerators totally ignore similar research in ecology that might lead to more efficient pest control, or healthier crops, or cleaner air and water. Even environmentalists veer away far too often from the practical to the emotional, getting sidetracked into single-issue campaigns such as saving wilderness or whales or Alaska and not paying nearly enough attention to building a world where life tables and the concept of environmental resistance are taught in the engineering schools right alongside multiplication tables and the concept of electromagnetic resistance—a world in which the savings of wilderness and whales and Alaska would be a normal part of everyday life, on a par with building a new skyscraper or designing a new computer circuit. Emotional commitment to nature is not enough. We need a revolution of practicality, and "balance of nature" needs to be more than a catchword—it must become one of the basic principals, like gravity and electricity and the second law of thermodynamics, around which we order our technology, our learning, and our lives.

It is important to emphasize that this is not so much a process of giving something up as it is of gaining something, or of replacing an incorrect—and therefore stultifying—theory with a correct one. As the whole host of products of organic chemistry that have so enriched our lives—the plastics, the synthetic fabrics, the linoleums, and all the rest—were not possible until chemists had abandoned the theory of phlogiston; as modern communications, radar, telemetry, and the like were not possible until physicists had put aside the concept of the aether; as America could not have been discovered until Christopher Columbus got it out of his head that the world was flat; so any real advances in the technology of life-manipulating disciplines such as agriculture and forestry must surely wait until we abandon our faulty concepts of the separability of life and replace them with the concept of *interdependence,* of looking at each species of plant or animal in terms of its function as well as its anatomy. For that is what the "balance of nature" is all about—interdependence. Interdependence, function, and connections. It is not, as it is popularly conceived to be, some mystical, unifying cosmic force-field; nor is it, as V. C. Wynne-Edwards maintained so passionately, some sort of "automatic mechanism" for population control that is innate in all animals and from which only humans are exempt. It is, instead, what Rachel Carson called a "state of adjustment." The forms of life have adjusted closely to each others' presence, over the eons of evolutionary time, and that adjustment has been very close to total, affecting all major aspects of a creature's life—what it eats, where it lives, how long it lives, and how often it gives birth. When humanity "controls" a species—that is, eliminates it, or greatly reduces its numbers—we are throwing those adjustments off, and doing it in a way which the ecologist Charles Elton once compared to flinging a crowbar into a delicate and complex piece of machinery. The machinery may keep running after the crowbar hits, but it is certain to be misadjusted, and it is no longer possible to predict what results it might produce.

It is, then, adjustments that this book is principally about. It is an attempt to show, using a real place and real events, what happens when humans throw those adjustments off and then fail to compensate. And it is a plea for understanding—for a realization of what we are doing, and a commitment to heed that realization, before we fling any more of Elton's crowbars into the system that has gotten along so very well without them for so many millions of years.

One of the most quotable of Rachel Carson's many pungent and well-turned observations about all this occurs almost at the end of *Silent Spring*. Discussing the concepts and practices of biological control—the techniques by which a pest organism's own natural enemies are turned against it, utilizing those adjustments we spoke of instead of proceeding as if they didn't exist—Carson writes:

> Through all these new, imaginative, and creative approaches to the problem of sharing our earth with other creatures there runs a constant theme, the awareness that we are dealing with life—with living populations and all their pressures and counterpressures, their surges and recessions. Only by taking account of such life forces and by cautiously seeking to guide them into channels favorable to ourselves can we hope to achieve a reasonable accommodation between . . . [them] and ourselves.

This, then, the "awareness that we are dealing with life," is the Carson Factor. As long as we take this factor into account, we can design with life, and our progress can be real. But when we ignore it—when we base our actions and our attitudes toward living things on the biological equivalent of phlogiston or aether or the flat-earth theory—then our progress will be false, and the unforeseen results of it will continue to come back and haunt us.

Awareness of life: An awareness implying respect for, knowledge about, and peace with. The Carson Factor. Will we be able to handle it? Or will our failure to grasp it, and all that it implies, mean that the results of our mistakes will continue to handle us?

Toward the first of these possibilities—and to the memory of Rachel Carson, who showed the way—this book is respectfully dedicated.

Acknowledgments

THIS BOOK COULD not have been written if a girl named
Melody James had not put aside her violin—on which she
won a scholarship to Whitman College—and decided to
major in biology instead. Her degree has enabled her to
help me compensate for my own abysmal lack of training
in this field, and the fact that her special field of interest
within biology has always been mammalian ecology has
been even more helpful. Besides, she's also an awful
lot of fun to be with—which, I suppose, is why we've
been happily married for eleven years now.

Besides my wife, the help of five other people has been
crucial. Dr. Steve Cross, professor of biology and small-
mammals specialist at Southern Oregon State College, did
much more than provide all the quotes to which his name
is attached in the text: He also gave me a quick course in
population biology, provided me with mountains of reading matter, critiqued the theories that Melody and I
developed to explain the Klamath irruption, and subjected
the paper I wrote explaining those theories to what is quite
possibly the most grueling test imaginable—he used it as

a partial text for two courses, one at SOSC and the other at the Oregon State University Field Station at Malheur National Wildlife Refuge. Walt Jendrzejewski, whose position in the county agent's office during the irruption enabled him to get an overview of the situation on a scale shared by few other people, gave me far more material than I used. Dayton Hyde cheerfully opened his home and his memory to me, even though he is working on a book about predator/prey relationships, which could easily be considered to be in direct competition with this one. Ed O'Neill, whose long tenure as staff biologist at Tule Lake Wildlife Refuge has given him an encyclopedic knowledge of the animal life of the Klamath Basin, filled in several key holes in my data—some of which he had poked in the first place—and provided an excellent sounding board on the dynamics of mouse irruptions while the theories in this book were under development. And finally, there was Paul Hatchett, whose expertise was invaluable on two levels: as a professional rodent-control man, and as a Klamath Basin rancher with firsthand knowledge of the problems caused by all those mice in 1957–58.

Beyond these six important people there have been many, many others, of whom it is possible to single out only a few. Rod Badger—chemistry professor, falconer, and enthusiastic amateur mammalogist—loaned me his copy of Adolph Murie's extremely hard-to-get book on coyotes in the Yellowstone, helped me locate information on synthetic pesticides, interpreted that information once I had found it, and came along on what would otherwise have been an awfully long, lonely drive when I went to interview Dayton Hyde. Rodney Todd, current Klamath County agent, was too new in the office to give me any direct help, but he did provide the information that led to

Paul Hatchett and Walt Jendrzejewski. Charles Funk gave
me several valuable leads from his position as chairman of
the Klamath Falls Group of the Sierra Club, and my old
friend Larry Chitwood, who was in high school in Klamath
Falls during the irruption, helped considerably to round
out my picture of spraying activities in the basin at that
time. The staff of the Southern Oregon State College
library in Ashland, where virtually all of my documentary
research was done, were extremely accommodating and
kind, especially in obtaining interlibrary loan materials
for me: Cliff Wolfsehr was particularly helpful in this
regard. Vern and Jean Crawford provided moral support,
expertise (both are naturalists) and some much-needed
baby-sitting. Max Gartenberg, besides acting as my agent
to find a publisher for the book in the first place, turned
himself into a small-scale clipping service for articles on
coyotes. Sandra Choron, my very capable editor at
Hawthorn, did all those essential things that only an
editor can do (by phone, by letter, and perhaps also a bit
by mental telepathy) to make an author turn out a man-
uscript and not just a pile of words—all of which have been
welcome, even if we *did* argue over the title for more than
four months. And lastly, let me thank the several current
and former citizens of Klamath County who gave me information on the mice but refused to let me use their names.
You all know who you are, and you know more than anyone else ever will how much of this book I owe to what
you told me.

IN SOME QUARTERS nowadays it is fashionable to dismiss the balance of nature as a state of affairs that prevailed in an earlier, simpler world—a state that has now been so thoroughly upset that we might as well forget it. Some find this a convenient assumption, but as a chart for a course of action it is highly dangerous. The balance of nature is not the same today as in Pleistocene times, but it is still there: a complex, precise, and highly integrated system of relationships between living things which cannot safely be ignored any more than the law of gravity can be defied with impunity by a man perched on the edge of a cliff. The balance of nature is not a *status quo;* i t is fluid, ever shifting, in a constant state of adjustment. Man, too, is part of this balance. Sometimes the balance is in his favor; sometimes—and all too often through his own activities— it is shifted to his disadvantage.

Rachel Carson, Silent Spring, 1962

The Carson Factor

Prologue

IT MIGHT BE BEST to begin with a bit of a fable.

Many centuries ago—a millenium, give or take a decade or two—there lived, in the town of Fulda, in the country of Germany, a certain abbot named Hatto, who desired riches and fame and power above all other things. And it came to pass, in the year of our Lord 968, that his wish was granted; for in that year the archbishopric at Mainz fell vacant with the death of William of Saxony, and Hatto was able to weasel his way into the position.

It was a pretty important position. Mainz, situated on the west bank of the Rhine opposite the Main River, was the ecclesiastical center of Germany and the most important focus of Catholicism in Europe outside of Rome itself. Under the name Germania Superior, it had been the capital of Germany during the Roman occupation, some seven centuries before; now with the birth of the Holy Roman Empire just six years previously, in 962, the ancient Roman traditions had gained a new popularity, and Mainz's connection with those traditions had given the city a glamour matched by few other places and made

1

the archbishop there one of the most powerful men in the world. For in those days religious and political power were inseparably intertwined, and an archbishop was a temporal leader at least as much as he was a spiritual one. Hatto had been made an abbot by the pope, but he had been appointed archbishop of Mainz by his friend and confidant King Otto I of Germany, known to history—for other reasons—as Otto the Great.

Hatto was, the legends tell us, not a particularly nice guy. Medieval politics was full of intrigues, counterintrigues, plots, murders, coups d'etat, and all sorts of other Machiavellian machinations, and Hatto was right there in the midst of them, intriguing, plotting, and murdering to beat the band—which was how he had got to be archbishop in the first place. He had friends in the nobility and he used them, scorning and disparaging the common folk while fawning on those who had money and power. He plotted with Otto against certain nobles; he plotted with certain other nobles against Otto. It is said that he strangled one opponent to death with his own priestly hands, using a golden chain to do the deed because the man was a member of the royal family. By these means and others Hatto soon amassed considerable wealth, owning, among other things, several of the largest and best-stocked granaries in the region and a tower on a rock in the middle of the Rhine at Bingen, twenty-five miles below Mainz, where the river leaves the fertile upper valley to enter a long gorge between the Hunsrück and Taunus mountains. The tower was an investment in security. It was tall, and strong, and formidably defended by rapids on three sides. Should any of the archbishop's many enemies decide to come for him, the tower would be a place to which he could retreat.

Against the enemy that finally came for him, however, the tower turned out to be no use at all.

The Carson Factor

The end came in the fall of 970, just two years after Hatto had been given the archbishopric. The harvest had failed, and there was great famine in the land; and Hatto, as the owner of all those well-stocked granaries, was constantly besieged by the common folk to share some of his bounty. He refused, angrily. The common folk had done nothing for him—what should he do for them? That grain was to be saved for his friends in the nobility. The legends have him holding out for a long time, and then at last, in an appearance of accession, telling the people to assemble on a certain day in a certain barn, "in order," one can hear him saying, "that there might be more grain for all."

The people assembled on the appointed day: men, women, and children, old and young, emaciated and starving, bearing on their backs the bags that they hoped the archbishop would fill with grain. Hatto herded them all into the barn; and when it was full, and could hold no more, he ordered the great doors closed and barred, struck a light, and set fire to the place. A great conflagration enveloped the barn and ascended to heaven, the roar of the flames mingling with the screams and cries of those being roasted to death within, while the archbishop walked about outside, warming himself by the flames—it was a cold day—and rubbing his hands with glee as he thought of the wonderful deed he had done. "They were like mice," he murmured to an assistant, "fit only to consume the corn. Now there will be more for the rich. Hark! Hear their screaming. Does it not sound like the piping of tiny mice?"

He stopped. There *were* mice in the building. They were pouring out of the burning structure from all sides, eyes shining, tails waving, and they were headed straight toward him.

The bishop fled; the mice pursued. There were hundreds

3

of thousands of them, perhaps millions, and they flowed over the ground like a wave lapping at his feet. He reached the Rhine. There was a boat there; he commandeered it, pointing it downstream toward Bingen. There his tower was waiting—his strong tower, his impregnable tower, made of stone and surrounded by seething rapids. Once inside the tower, he would be safe. Surely the mice couldn't reach him there!

But the mice—and the wrath of God that drove them—had other ideas.

What happened next has been told in a variety of ways. The best-known account is probably that set down by Robert Southey, the English doggerelist, in his *Legend of Bishop Hatto,* composed in the early years of the nineteenth century. Southey wrote:

For they have swum over the river so deep,
  And they have climbed the shore so steep;
And up the tower their way is bent,
  To do the work for which they were sent.

They are not to be told by the dozen or score,
  By thousands they come, and by myriads and more;
Such numbers had never been heard of before,
  Such a judgement had never been witnessed of yore.

Down on his knees the Bishop fell,
  And faster and faster his beads did tell,
As, louder and louder drawing near
  The gnawing of their teeth he could hear.

And in at the windows and in at the door,
  And through the walls, helter-skelter they pour,
And down from the ceiling and up through the floor,

From the right and the left, from behind and before,
And all at once to the Bishop they go.

They have whetted their teeth against the stones,
    And now they pick the Bishop's bones:
They gnawed the flesh from every limb,
    For they were sent to do judgement on him!

Or, to use the words of the early seventeenth-century English writer Thomas Coryat, in his *Crudites,* "the prelate retreated to a tower in the Rhine . . . but the mice chased him continually . . . and at last he was most miserably devoured."

So much for the legend. What about the facts?

Hatto himself was real enough; there are contemporary records, scanty but apparently accurate, of his twenty-five-year abbothood at Fulda and of his two brief years as archbishop of Mainz. The portrayal of him as cruel and grasping is also apparently accurate, although some of the atrocities attributed to him may actually have been committed by a predecessor, also named Hatto, who reigned at Mainz earlier in the same century and is said to have met his end in quite as romantic a fashion, tossed into the crater of Mount Etna during one of its eruptions by a party of his disgruntled subjects. The tower is real, too: in fact, it still stands on its rock in the river just below Bingen, where is is known as the *Mäusethurm,* or Mouse Tower. However, at about this point we run out of truth. Records show clearly—alas for fable!—that the tower never belonged to Hatto. It could not, for it didn't exist in his day. The tower was built in the eleventh century—not the tenth—by one Bishop Siegfried, as a tollhouse to collect tariffs from barge traffic on the Rhine. The German word for toll is *Mauth,* so the tower was called the *Mauth-thurm.* The eventual conversion of this name to *Mäuse-thurm,* and its attachment to the legend of Bishop Hatto when it arose early in the thirteenth century, was probably inevitable. And as for the mice themselves—well, the Germans of that day had a peculiar belief about mice. They thought they represented the souls of the dead. The mice pouring out of the bishop's burning barn were obviously the souls of those the bishop was merrily roasting to death inside, and their one purpose was to avenge the demise of their former owners. Actually, Bishop Hatto's was not an isolated case: There are many tales in Teutonic folklore of evil characters being devoured by mice, and most of them parallel the tale of the bishop exactly, right down to the burning barn and the tower in the river. E. Cobham Brewer's *Dictionary of Phrase and Fable* lists at least four others besides Hatto, including a Bishop Widerolf of Strausburg—eaten alive for having destroyed a convent— and one Count Graaf, who is interesting because the tower where the mice got him is explicitly called a *toll* tower. It was evidently quite a popular way to go. Besides, the very idea of mice assembling in such large numbers, and doing such great damage, is laughable, isn't it? Perhaps the old German chroniclers were stretching things a bit?

Perhaps. And perhaps not. Let us now make a great leap, forward almost one thousand years and westward several thousand miles, across an ocean, a continent, and several ages, to western North America one winter not so very long ago.

1

Wee, sleekit, cow'rin, tim'rous beastie...!
Robert Burns, *To a Mouse*, 1785

O N SUNDAY, January 20, 1957, as General Dwight
David Eisenhower was being inaugurated for his
second term as president of the United States, two feet of
snow fell on the city of Klamath Falls, Oregon, bringing
an end to the longest winter dry spell in the region's
hundred-year history. The last rain of any consequence
had fallen in October; since then, the skies had been clear,
the days had been bright and warm, and the local farm
authorities had been worried. Agriculture in the vicinity
of Klamath Falls depends almost entirely on irrigation,
and irrigation depends on a good snowpack in the surrounding mountains. That snowpack was practically
nonexistent. In nearby Crater Lake National Park, where
the snow should have been eight feet deep, there were
barely twenty-six inches; the Sun Mountain snow course
east of the park had less than a foot, and at the Taylor
Butte snow course in the northeastern part of Klamath
County there were only scattered patches a few inches
thick. The weather bureau hydrologists sent out to measure the two courses on the last day of December had

7

come back shaking their heads. They had taken snowshoes along, but it had been wasted effort. Pickups had been able to reach both courses with ease.

In the Klamath County office of the Federal Cooperative Extension Service, on the third floor of the old post office building at the corner of Seventh and Oak in downtown Klamath Falls, the drought had caused a noticeable increase in the burden of the already harried staff. County Agent Charles A. Henderson and his assistants were besieged by concerned farmers wanting advice on ways to get through the dry spell. How would these abnormal conditions affect sowing times and the amount of acreage devoted to each crop? Were there drought-resistant seed varieties they should be trying? Would modification of irrigation patterns be necessary? Would crop rotations be upset? The questions continued to come in; the agents continued to answer them, and to fall further and further behind their normal schedules.

And then, toward the middle of the second week in January, a strange new complaint began to show up. Something was eating the bark off young fruit trees: They were being found at scattered locations about the basin, girdled to a height of five to six inches, dying or dead, with tiny tooth marks on the bare sapwood. An animal, obviously— but what kind of animal was it, and how could it be combatted? Fortunately, the overworked agents saw at once, this was not a problem they were going to have to deal with personally. There was someone else they could hand the problem to, someone conveniently nearby. "The county had a Weed and Rodent Supervisor," recalls Walt Jendrzejewski, one of Henderson's assistants during that period, "and it was customary to quarter him with us." This weed and rodent supervisor, sixty-year-old Harold Schieferstein, was obviously the one to handle the problem of the bark-eaten fruit trees. The county agents pointed the fruit tree owners in Schieferstein's direction and went back to the problems caused by the drought.

Schieferstein had no trouble identifying the culprits; they were, as the county agents had also known and most of the farmers had suspected, voles—field mice—of the variety known scientifically as *Microtus montanus* and called by local citizens "Klamath mice," "tule mice," or "marsh mice." It wasn't too difficult to prescribe a cure, either: The normal mouse bait—hulled oats laced with strychnine—ought to work. What made the situation unusual, Schieferstein realized, was its timing. Mouse damage of this type is normal under a heavy snow cover, but it is almost unheard of during a dry winter like the one the county was currently experiencing. Perhaps it was worth a newspaper article. And if the article were written carefully, it might serve the additional purpose of giving the farmers enough information so that they wouldn't have to call the county agent's office for advice. This would allow Schieferstein and the agents to forget about the mice and concentrate on the more important problems raised by the drought.

Schieferstein's article appeared in the local newspaper on January 20, an unfortunate accident of scheduling that promptly relegated it to obscurity behind the big news of the snowstorm and Eisenhower's inauguration. That was a pity. It was a very nice article—four column-inches of good, pithy, down-to-earth, matter-of-fact information and advice. The weed and rodent supervisor had done his work well, and under normal conditions this would have been the end of the story. These were, however, far from normal conditions. Schieferstein had no way of knowing it, of course, but his little article marked almost the last time anyone in Klamath County could afford to be matter-of-fact about mice for quite a long time to come.

2

KLAMATH COUNTY is the kind of place that nearly everyone would like to be able to call home. More than six thousand sprawling square miles of forests, lakes, green fields, tall mountains, rushing rivers, and clear air—incredibly clear air—it is the most prosperous county in eastern Oregon and one of the wealthiest rural areas in the western United States. Its western boundary wanders along the crest of the Cascade Mountains for more than one hundred miles; its southern edge butts up against California, running the forty-second parallel eastward from the Cascades better than halfway to the Nevada line across the green ridges of the Warner and Gearhart ranges. Most of this vast expanse of territory—an area larger than Connecticut and Rhode Island put together—is timberland, a jumble of rolling hills and steep, rugged mountains covered with dense forests of Douglas fir, true fir, larch, and lodgepole and ponderosa pine. The rivers foam through narrow and crooked canyons; the ridges march haphazardly away to all points of the compass, and vision in any direction seldom exceeds a mile. It is no country for claustrophobes, but for anyone else it is close to paradise.

The Carson Factor

The one great exception to this pattern is a broad, steep-sided, roughly circular valley in the extreme southern portion of the county—so far south, in fact, that although it is almost always thought of as being part of Oregon, nearly half of its thousand-square-mile floor lies over the border in California. Though subdivided by an accident of civil geography into two states and three counties, the great valley operates—ecologically, agriculturally, and to a large extent politically—as a single unit. It is called the Klamath Basin.

The floor of the Klamath Basin lies at an elevation of nearly four-fifths of a mile above the Pacific Ocean, but because the surrounding mountains shoot up to as much as two and three times that height, this elevation is not easily comprehended. The basin is forty-two degrees above the Equator and one hundred forty miles inland from the sea, which puts it in about the same relationship to the west coast of the continent that the upper Hudson Valley bears to the East Coast. The landscape, however, is anything but Hudson-like. This is some of the youngest, rawest land in North America. Cinder cones and fault-block mountains ring the basin and thrust up from its flat green floor like sentinels. To the south looms the vast, other-worldly bulk of volcanic Mount Shasta, described long ago by the poet Joaquin Miller as "Lonely as God, white as a winter moon," its sides scoured by living glaciers, its summit fumaroles still steaming. To the north squats the mangled stub of Mount Mazama, which blew up at about the time the pyramids of Egypt were being constructed, scattering pieces of its summit as far away as Canada and creating that magnificent, cliff-ringed body of achingly blue water we know today as Crater Lake. In the basin itself, tucked up in the southwest corner beside wide, shallow Tule Lake, Lava Beds National Monument protects nearly fifty thousand acres of cinders and basalt that poured northward out of the Medicine Lake Highlands as recently as the time of Columbus. But perhaps the most striking sign that the fires that shaped this land are still alive lies within the basin's population center, the city of Klamath Falls itself. Driving about the city you will see, at odd intervals, small pipes stuck in the ground from which steam bellows forth in billowing, unending clouds, as if it were coming from the earth itself. It is. Klamath Falls is built squarely on top of a subterranean lobe of magma, and a substantial number of the city's homes and businesses are heated by geothermal energy.

As of 1978, the official population of Klamath Falls was only 17,285, which would make it seem nothing much more than an overgrown village. This figure, however, is very misleading. The city boundaries are quite tightly drawn, and nearly two-thirds of the urban area lies outside them, bringing the effective population up to something closer to fifty thousand—fifty times greater than anything else within a radius of more than sixty miles. There is a college there (Oregon Institute of Technology); a ninety thousand-volume public library; a good school system; an active community concert association; and a remarkably fine daily newspaper called the *Klamath Falls Herald and News*. The city occupies a splendid setting on a group of low hills clustered at the southern tip of twenty-mile-long Upper Klamath Lake. To the west loom the high, timbered ridges and snowcapped volcanoes of the Cascades; to the north, dotted with islands, spreads the 142-square-mile surface of the lake, largest in Oregon; and to the east and south, broad and green in the mountain sunlight, stretches the principal focus of our story—the rich, flat fields of the basin. They are fields to be proud of. This is quite possibly the finest agricultural land in the United States, with per-acre crop production running as high as twelve times the national average. Less than one-twentieth of the basin's three hundred thousand acres of irrigated cropland is devoted to potatoes, for example, but this tiny acreage produces more than 15 percent of the nation's sixteen million annual tons. Even less acreage is planted to clover, but the basin's yearly alsike clover seed harvest is 4.5 million pounds, enough to account for nearly a quarter of U.S. production and to earn the basin the self-proclaimed title of "alsike clover seed capital of the world." Oats, wheat, and barley are also grown here, along with small amounts of orchard products and other vegetable crops; but all of these things, including the highly visible potato industry, are relatively minor components of the basin's economy, which runs principally on lumber products and livestock. The lumber products come down out of the nearby mountains; the livestock, largely cattle and sheep, is raised in the basin itself, and most of the agricultural land is devoted to pasturage and to hay crops such as alfalfa and bluestem grasses.

One of the best ways to get a feel for the wealth of this area is to drive State Highway 39 south out of Klamath Falls, through Merrill, Tulelake, and Stronghold, and eventually, over the mountains and some two hundred fifty miles down the pike, into Reno. For mile after mile the fields stretch out on either side, smooth and level, tawny with harvest in the fall, rich-loam black in the winter, and in the spring green with the miracle of growth. Fat cattle graze slowly in the fields; houses and barns wear a prosperous and contented look. There is nothing to suggest that it was ever any different. You cannot see the trauma the land has gone through, nor the damage, long since healed, that once marked the pastures and fields and lawns of this beautiful land as the legend of Bishop Hatto came to life and a creature we all know as a minor nuisance became, for all of one all-too-lengthy winter, a living nightmare.

3

ONE BRIGHT MARCH Monday in 1978, a little more than twenty-one years after Harold Schieferstein's mouse warning had been upstaged by Dwight Eisenhower and a blizzard, I drove into the Klamath Basin from the west, winding down out of the Cascades on State Route 66—known locally as The Greensprings—and arrowing across the broad flat fields near Keno, on my way to Klamath Falls and the heart of a mystery. A few months previously, leafing through an old wildlife magazine in my den, I had chanced upon a reference to the "largest population explosion of meadow mice in world history, save for one," said by the author to have taken place in Klamath County some time in the 1950s; and I had been immediately intrigued. I thought I knew the Klamath country well—I have close friends there, have visited it many times, and have lived just over sixty-five miles west of Klamath Falls for the past eight years. But the fact that mice—mice!—had once brought the county to its knees, and not so very long ago, was something that had escaped me. Why was this not better known? What had the basin been like during the great mouse explosion, and what had caused the thing in the first place? It was this last question that especially intrigued me. If it were true, as the author of the article I was reading maintained, that this disaster had been brought on the people of Klamath County by massive predator-control programs that had killed off the coyote population, then the story ought to serve as a splendid example of how mistreating nature brings its own punishment, and should be extremely valuable for that reason. On the other hand, if it were *not* true—if the coyotes had had little or nothing to do with the explosion of mice—that would be valuable too. The issue of predator control versus wildlife preservation has become entirely too emotional in recent years, and one of its fiercest battlegrounds has been the question of whether or not predators do any good. Livestock interests have always claimed loudly that they do not; environmentalists have maintained just as loudly that they do. And because both sides are guilty of using their views on the matter as crutches to support preconcieved opinions (the livestock people, that the coyote should be eliminated; the environmentalists, that it should be left alone), neither appears really to want to know which view is right. It is past time—long past time—for the argument to leave pyrotechnics behind and get onto a firm basis of fact. This Klamath situation looked like a good place to begin.

I decided to try to find out more about it.

It wasn't going to be easy. I found that out almost immediately, running quickly into a series of stone walls. The facts were there, all right: Copies of the local newspapers for that period were available on microfilm, and information from them had rapidly filled a fat notebook. But scholarly interpretation of those facts was another matter. There was one publication, put together by scientists who had been involved in a control effort during the mouse plague; it had been published less than six months later, and it was long on detail but short on perspective. Other than that, there seemed to be nothing. No Ph.D. thesis; no seminar reports; not even, as far as I could tell, a student research paper. It was maddening. Here was one of the greatest agricultural disasters in the history of the United States, a disaster costing the affected farmers millions of dollars in damages, a disaster almost mythical in nature—remember the Pied Piper?—and scientists had apparently paid no more attention to it than if it had taken place on the far side of the moon. It was going to take a lot more than just the facts from the newspapers to be able to prove anything. Which was why I was now headed, on a bright morning with the birds singing and a feeling in the air as though the only thing in the world that made any sense would be to lie back on a rock and soak up some sun, toward a succession of dark, dusty offices in Klamath Falls.

Perhaps that is painting too bleak a picture. There were certain things that were known about those Klamath mice, among them something that, to me, seemed quite startling. Driving eastward now, across the western arm of the basin and toward the city, I thought of that fact. Those "mice"— it had become apparent quite early on—were not, strictly speaking, mice at all!

To be sure, it is difficult at first glance for the casual observer to notice much difference. Like a mouse, the creature that ate Klamath Falls is small and brown, with a habit of scampering rapidly over the ground, a voice that squeals loudly in distress, and a large appetite for plant matter, often satisfied at the expense of the vegetable garden or the pantry. It is no wonder that the farmers fighting the beasts, the reporters writing about them, and

17

even the scientists studying them, usually referred to them as mice. But the differences are there; it is merely a matter of looking for them.

It was Dr. Steve Cross who had told me about those differences. Cross is a biology professor at Southern Oregon State College in Ashland, sixty-five miles west of Klamath Falls across the Cascade Mountains. He is a tall, slender, balding man, on the young side of middle age, with a heavy black beard that gives him a startling resemblance to the character called Oscar Boom in the "Alley Oop" comic strip—except that Boom rarely smiles, and Cross smiles a great deal. He was smiling as we spoke, obviously amused at my request to explain the difference between mice and field mice. "It's always been difficult for me to describe the difference to my students," he chuckled. "I guess you could say they look like a real mouse, only fuzzier. Just a little ball of fur with eyes. Come on, I'll show you."

He led me into the laboratory next to his office, pausing at a set of gray metal filing cabinets ranged along one wall. "This cabinet," he said, gesturing at one, "contains our teaching collection of mammals found in this area. Now, look." He pulled open a couple of drawers, and I saw that they were full of neatly stuffed and labeled dead animals. One contained mice, familiar-looking little creatures with silky fur, long tails, pointed and inquisitive-looking noses, and the big, prominent ears that were the model for those big round flaps you see on children's Mouseketeer hats. The second appeared to contain mice, too, but a closer look belied that first impression. The bodies were larger and chunkier; the noses were blunter, the fur rougher, the legs and tails shorter and stubbier, the ears next to non-existent. It was difficult to tell about the eyes—they were just holes in the skin with the cotton stuffing showing through—but they gave a strong impression of being quite

18

small for the size of the animal. Not mice, then: but if not mice, what? Cross supplied the answer over my shoulder. "Voles," he said.

Voles. The word seems faintly archaic and Middle-Englishy, as if Chaucer's pilgrims might have encountered the creature on the road to Canterbury. They might well have done so, too, but that doesn't mean that it's something exotic. On the contrary: Voles are among the most common mammals on the face of the earth. They are found all over the northern hemisphere, from above the Arctic Circle almost to the Equator, on every continent and nearly every island. Most of us call them field mice, or meadow mice, and that is not really wrong—they have been called that for millenia, and the word "vole" itself comes down to us, through several derivative steps, from the Old Norse *vollmus,* which translates directly as "field mouse." But it is not really right, either. Gazing rather abstractedly at his two filing drawers full of animals, Cross searched for a way to explain why. "I guess we'd better start at the beginning," he said finally.

"As I said, this cabinet contains our teaching collection of mammals. Now, you'll notice that the *Rodentia* start here, about halfway up this side, and go clear to the bottom of the other side. They're by far the largest order in this area, and the voles are the largest family in that order. They belong to the family *Cricetidae,* which is divided into two subfamilies, the *Cricetini* and the *Microtini.* The *Cricetini* are what you usually think of as mice; they include the harvest mice, fulvous mice, deer mice, and so forth. The *Microtini* include the voles and lemmings." So we are not talking about merely different species, or even different genera, but different *groups of genera*—a rather far piece from each other on the mammalian family tree, even if they are both commonly called mice.

Their food preferences and living habits are nearly as

different as their physiology. Field mice are grass-eaters; "true" mice, seed-eaters. Field mice may be seen at any time of the day or night, whereas true mice are almost entirely nocturnal. Field mice are what Cross calls "very runway-oriented": They build meticulously maintained paths ("runways") through the grass, along which they dash in a manner that one zoologist, Dr. Robert M. Storm of Oregon State University, has compared to a slot-car racer dashing along its track. True mice make no runways, and often don't live in the grass at all. But the most important difference between them, for the sake of our story, has nothing to do with their physiology or with their behavior as individuals. It has to do with their lives as a group. Populations of true mice, it turns out, are not strongly cyclic. Populations of field mice are.

Populations of field mice, together with those of their close cousins the Arctic lemmings, seem to be bound to an innate three- or four-year rhythm of boom and bust. Driven by some mysterious mechanism that scientists have yet to understand completely, these populations grow and shrink, shrink and grow, expanding and contracting on a scale that amounts to a several hundredfold difference in density between highs and lows, with a regularity approaching that of a clock. At the low point of the cycle, an acre of grassland may hold between one and two voles; at the high point, the same acre may be crawling with as many as four hundred. The time from high point to high point, or from low point to low point, is usually three years, but it may be as few as two or as many as six. Graphed, the cycles look like a sawtooth wave, with a slow increase followed by an abrupt decrease. The size and shape of these waves remains remarkably constant over long periods of time.

What causes these cycles? Biologists would give much to know. Later on in this book we will be discussing several theories: Here, we should keep just three points in mind. First, these field mouse population cycles are natural; they are not induced by the activities of man, but result either from a genetic quirk in the makeup of the mouse, or from some interaction of the mouse with its environment, or possibly both. Second, these causes are sufficiently complex that they are not yet understood. And third—*they could not have been the primary cause of the great mouse explosion of Klamath County.* These incidents could not have happened, it is true, *without* the cycles: There had to be a tendency toward rapid population growth, as represented by the increase phase of the cycle, before that growth could occur. But, by the same token, the fact that these increases occur regularly every three or four years without, normally, causing anything more serious than a momentary lull in the cat-food market, indicates that something was different about the problems experienced by the farmers of Klamath County during the winter of 1957–58. There must have been some imbalance in the forces regulating the field-mouse cycle, allowing the cycle to get out of hand and grow far, far beyond its normal collapse point. What was that imbalance? What caused it? Why had it not occurred before, and why hasn't it happened since? These are the questions I wanted answers to, answers I hoped might lie somewhere in the tangled course of events of that Winter of the Mice. And that was why I had come to Klamath Falls—to get as clear a view as possible of those events, and of the milieu in which they took place.

I drove down the long hill above town, across the Link River and past Lake Ewauna, and into the city.

4

THE RECEPTIONIST IN the county agent's office wanted to know what my business with Rodney Todd was, so I told her: "mice." An old farmer standing a few feet away snorted loudly. "Don't tell me," he exclaimed, rolling his eyes heavenward, "that we've got *mice* again!" I assured him that I was there to talk about the mice he had had in the past. He sighed with relief. "Thank God." he said. He would not elaborate further. It was an attitude I was to run into often over the next few months as I probed for details of what is known officially as the Oregon Meadow Mouse Irruption of 1957. Mice remain, to this day, something Klamath County would prefer to forget.

Rodney Todd showed up a few minutes later. He is a big, blond, vital young man with a pleasant grin and an easygoing manner that appears to hide a vast reservoir of potential energy. We ambled down the hall to the office he occupies as current Klamath County extension agent, and he plopped down behind an ancient oak desk. "I probably can't help you very much," he said apologetically. "I've only been here a few years." Most of the old-timers, it seemed, were gone. He thought Henderson and Schieferstein were both dead. I mentioned a few more names, gleaned from hours of research in the back files of the *Herald and News*. J. D. Vertrees? He didn't know. Tom Horn? It didn't ring a bell. Walt Jendrz- Jendrzj- J-

Todd brightened. "Jenn-*jess*-kee," he said. "You ignore all those other letters. Yes, Walt Jendrzejewski is alive and well and selling real estate in Klamath Falls. He can probably give you quite a bit of information. Here, I'll give you his number." He scribbled something on a piece of paper and handed it across the desk. We exchanged a few more pleasantries and parted. Todd went down the hall to help the man who had exclaimed about the mice; I headed for the nearest phone booth.

Two days later I was back in Klamath Falls, settling down to a cup of coffee in the family room of Walt Jendrzejewski's big hillside home in the affluent northeastern section of the city. Large windows looked out over houses and businesses to distant Lake Ewauna, gray under leaden March skies. "I don't know if I can help you or not," he had told me over the phone, "but I'm sure willing to try." Now he sat in the armchair opposite me—a short, barrel-chested figure with close-cropped black hair and watery eyes that seemed to be squinting into a perpetual wind, as if he had just come in from riding fence. He looked the very epitome of a contemporary western small rancher, a look that was undoubtedly extremely helpful to him in his years of county-agenting. I liked him immediately.

He sipped slowly at his coffee. "I don't know if anyone knows where they all came from," he said after a while. "Of course, these things are natural here, they live in the marshes. And a lot of the marshes have been drained and put mostly into grass, but these things are a problem every

24

year in conditions like that, when you've got subirrigated land in sod. Now, their method of combatting them had developed over a period of time, to use the winter flooding period to drive 'em up on the banks and bait 'em there, and we got pretty good control that way. Most years, that is. Nineteen fifty-seven was different. But this lasted more than one year, you know, it built up over a period of time, and people didn't really notice it at first. You know, a farmer's got a lot to do, he doesn't get out in his fields too often. I don't think the full extent of damage was realized until they began running machinery over it. . . ."

The machinery was still mostly in the sheds when the snow fell and the drought broke, that late January Sunday of 1957. Most of it would remain there for quite a long time to come. Some of it was being repaired: Winter is a traditional time for overhauling farm machinery, which never seems to work exactly as it was designed to, and the ruddy-cheeked ploughboy of bucolic legend often spends his winters looking more like a grease monkey than a husbandman. The rest of it simply wasn't needed yet. It was too early for plowing and planting; those farmers that weren't on their backs under their tractors and combines, or staring through welding masks at broken plow blades, were busy at their desks, working out planting patterns, ordering seed and fertilizer, and balancing their books from the previous harvest. There was some spraying being done, mostly soil sterilants such as chlorate-bordate or CMU, but this was commonly done by air, and the pilots would have very little chance to look for field-mouse runways. Other than this, the fields were left pretty much alone.

The drought was truly over: February came through wetter and warmer than usual, and the March snowfall at

Crater Lake broke all previous records. However, there was still cause to be worried about the summer. Even with all that recent snow on it, the total snowpack was still running only about 76 percent of normal. That snowpack was the reservoir from which irrigation water would have to come. Would there be enough of it, even yet? Would the magnificent ditch system of the Klamath country, one of the most successful reclamation stories in the history of agriculture, fail to deliver?

Irrigation has been a way of life in this land of little rain ever since the first settlers arrived, shortly before the time of the Civil War. It was not until Teddy Roosevelt and his Reclamation Act came along, however, that this way of life was elevated to a complete dependency which would alter the entire landscape of the basin. Back in those days, in 1902 and before, the Klamath country looked very different than it does now. Most of it was under water. Three great lakes—Upper Klamath to the north, Lower Klamath to the southwest, and Tule to the southeast—spread their shallow waters over a total of nearly three hundred thousand acres. Around the lakes were extensive areas of cattail marsh, providing sheltering, breeding, and feeding grounds for huge numbers of waterfowl. A number of smaller lakes—Swan, Round, Long, Short, Aspen, Agency, Ewauna, and others—lurked in various corners. The city of Klamath Falls, on the Link River in the north central part of the basin, had a population of twenty-five hundred, a rainfall of scarcely thirteen inches a year, and two steamship lines. One operated northward, across Upper Klamath Lake; the other operated southward, connecting via Lake Ewauna, the Klamath River, the Klamath Strait, and Lower Klamath Lake, to the railroad terminal at Dorris, California, better than twenty miles away.

Teddy Roosevelt changed all that. He had been a rancher

in the Dakota Territory and he knew how important water was to the arid West; and in his very first address to Congress, shortly after his abrupt elevation to the presidency following the assassination of William McKinley, he gave notice that government water policy was going to be a top concern of his administration. "Water and forestry reform," he emphasized, his high, reedy voice braying out over the assembled senators and congressmen, "are perhaps the most vital internal problems of the United States." Within hours of that message, a committee of senators had been formed to draft a reclamation bill, and within six months the bill was passed and signed into law. Almost immediately, teams of engineers from the U.S. Geological Survey began fanning out through the West, looking for places where their infant Reclamation Service could be of use; and one of the first spots they came to was the Klamath Basin.

There was already a fair amount of land under irrigation in the Klamath, for the simple reason that the low rainfall here makes farming absolutely impossible any other way. Several thousand acres of cropland and pasture were watered by ditches in the Swan Lake and Lost River sub-basins; and the Ankeny Canal—dug by a group of citizens called the Linkville Water Ditch Company back in the days when Klamath Falls had been known as Linkville, but now owned and operated by one Henry Ankeny—tapped the Link River just as it left Upper Klamath Lake and used its water to irrigate town lots in Klamath Falls and farms on the dry flats between Lake Ewauna and the Lost River. But these projects were almost unnoticeably small beside the Reclamation Service plan, which proposed nothing less than a complete rearrangement of drainage patterns and landforms within the basin. Two of the three major lakes, Tule and Lower Klamath, would be drained and

27

turned into farmland; the Lost River would be diverted into the Klamath, and most of the Link would be fed into a widened and deepened Ankeny Canal to irrigate the reclaimed lake beds. The project would affect nearly three hundred fifty thousand acres, take several decades to complete, and cost, back in the days when a dollar was a dollar, nearly $4.5 million.

The first step in carrying out the Klamath Water Plan, as it was called, was to gain control of the key Ankeny Canal. In this, Reclamation was opposed by something called the Klamath Canal Company, a group of local businessmen who believed ardently that the government should keep its bureaucratic nose out of their affairs, and that the basin would be far better off if it were developed by private enterprise instead. Both sides sent negotiation teams to discuss the sale of the canal with Henry Ankeny, and we probably have the incompetence of the Klamath Canal Company team to thank for the fact that the Klamath Project exists today. The single comic-opera meeting of that team with Ankeny took place in the canal company offices in downtown Klamath Falls, where it was recorded for posterity by an engineer named Don Zumwalt. There was apparently no small talk. Ankeny came in and sat down at a table opposite the three canal company officers, Charles N. Hawkins, William K. Brown, and Ben Gould. Hawkins spoke first. "Well, Henry," he asked jovially, "how much do you want for your ditch?"

"Twenty-five thousand dollars," said Ankeny, bluntly.

"Oh, but that's too much!" blurted Hawkins. Ankeny, obviously nettled by Hawkins's tone, rose from the table. "That's my price, gentlemen," he said. "If you don't want the canal, you don't have to buy it. Good day. " He walked out. There was a long silence, broken at last by an explosive sigh from Gould.

"Well, Charlie," he allowed, "You're pretty much of a damned fool, ain'tcha?" And thus the canal was sold to the government, which was then able to begin the next stage: the construction of the Lost River Diversion Channel and the drying up of most of Tule Lake.

To understand how this next stage worked, and how it affected the ecology—and the mice—of the Klamath Basin, we must first understand the rather peculiar drainage patterns involved. The Lost River, which is named for its habit of disappearing underground at odd intervals in the upper part of its course, begins in a body of water called Clear Lake, perched in the California mountains a few miles east of the basin and some four hundred feet above it. From the lake, the river flows northward into Oregon, looping through the fertile Poe and Langell valleys before turning westward to pour through spectacular Lost River Gap, rush past the tiny town of Olene, and burst into the basin proper. In the recent geological past, when the entire Klamath Basin was a single large lake, the Lost River ended here, mingling its waters with those of the lake and, through it, draining to the sea through the ancestral Klamath River on the far side of the basin. However, as the eons passed, this lake slowly silted in and dried up, until by the time the first humans arrived some fifteen thousand years ago there was nothing left of it but a series of gargantuan puddles—the historical lakes of the Klamath—scattered about the basin floor. During this drying-up period, one of the lakes—Tule—had become separated from the others by a low swell of silt, and now occupied a basin with no outlet. The Lost River flowed into Tule Lake, its annual discharge exactly matched by the lake's annual evaporation. This Lost River-Tule Lake system was closed: Despite the fact that it was separated from the Klamath by barely six miles of what appeared to be completely level ground, its waters never reached the sea.

To the engineers of Teddy Roosevelt's Reclamation Service, this seemed an ideal setup for exercising their particular talents. All they would have to do would be to build a large canal across that six miles of flat land and divert the flow of the Lost River into it. Deprived of its source of water, Tule Lake would dry up, releasing its bed for farmland. The old Lost River channel would become a canal, with a controlled flow of water in it; the deepest section of the old lake bed could be retained as a sump for irrigation runoff. Thousands of new acres of crops could go in, and the wealth of the Klamath Basin would be greatly enhanced.

When they surveyed for the canal they discovered that the Klamath River was slightly higher than the Lost River at their closest point of approach, meaning that the canal would run the wrong way, and by the time they had located a route that would guarantee that the water would always run outward the ditch length had stretched to eleven miles; but this was of small import. Excavation got underway early in 1911. In June, 1912, the Lost River Diversion Channel was complete and in operation, and Tule Lake had obediently begun to shrink. By 1923 it was down to two thousand acres and the land rush had begun. Soon, two towns—Tulelake, California, and Malin, Oregon—had sprung up on what used to be lake bed. At about the same time, the farmers discovered that Tule Lake soil was extraordinarily well adapted to growing potatoes. Soon nearly twenty thousand acres of the old lake were being planted annually to a crop whose potato would eventually become known as the "Klamath Gem." The Klamath Basin was on its way to becoming one of the potato capitals of the world.

In the meantime, at Lower Klamath Lake, a few miles west of Tule across an abrupt range of steep dry hills known as Sheepy Ridge, a similar drainage scheme was moving forward—although the Reclamation Service was not at first involved. It is quite probable that it wanted to be, but it had a conflict: Another of Teddy Roosevelt's pet projects was getting in its way.

The extensive marshes surrounding Lower Klamath were the largest for several hundred miles in any direction, and as such were of critical importance as a resting place for waterfowl following the Pacific Flyway, that great migratory route used by all birds west of the Rocky Mountains. Each spring and fall the sky over the Klamath country was black with literally millions of ducks, geese, and swans. It was a sportsman's paradise, and Teddy Roosevelt was a sportsman. In 1907, by presidential proclamation, he had set aside nearly half of Lower Klamath Lake and its surrounding marshlands as the nation's first waterfowl refuge. That meant that the birds would have to take priority over the farmers. The Reclamation Service—perhaps reluctantly, perhaps not—declared a hands-off policy.

Unfortunately, this policy was not written into law, and it failed to impress a group of citizens who called themselves the Lower Klamath Drainage Association. To these people, the marshes remaining outside the refuge—critical waterfowl habitat or no—represented land going to waste, and they were determined to put them into production. If Reclamation wouldn't help them, they would just have to proceed on their own.

Lower Klamath Lake, like Tule Lake, was part of a rather peculiar drainage system. The key components of that system were a mile-long watercourse known as the Klamath Strait and a natural basalt dam across the Klamath River, a few miles downstream from the mouth

31

of the strait, called the Keno Reef. The strait was dead
level and could flow in either direction. It connected the
lake to the river. At times of high water, the river's flow,
backed up behind the Keno Reef, would pour eastward
through the strait and fill up the lake; at times of low
water, the strait would reverse, with water pouring westward from the lake to augment the flow of the river. This
made it a superb natural flood-control facility, but the
Lower Klamath Drainage Association couldn't see that
any more than they could see the need for the waterfowl
refuge. All they saw was the vulnerability of those two key
points, the strait and the reef. If the reef were breached,
the level of the river would be permanently lowered so
that the strait would flow only outward, and the lake
would drain; alternatively, if the strait were blocked, the
inflow of water from the river would cease, and the lake
would dry up. The only real question was which of these
two attractive methods should be used—and while they
were arguing that point, the Southern Pacific Railroad
came along and solved it for them. In 1907, the same year
that the Lower Klamath Wildlife Refuge was created, the
railroad began construction of an earthen causeway
across the west end of Lower Klamath Lake. The causeway was completed two years later. It included a massive
culvert at the Klamath Strait that was designed to allow
the strait to continue functioning, but the culvert had
gates and the gates could be closed. The Lower Klamath
Drainage Association stepped up its pressure on the
Reclamation Service—now the United States Bureau of
Reclamation—and in less than a decade, they won. On
November 30, 1917, the bureau and the railroad signed
a pact calling for permanent closure of the Klamath
Strait gates, and by 1922 the lake was totally dry. Not
even a puddle remained.

Now, it's difficult to imagine a more useless piece of real estate than a waterfowl refuge without water, and the supporters of the Lower Klamath Wildlife Refuge were, quite naturally, furious. The furor was calmed a bit in 1928 by the dedication of two more Klamath Basin refuges, at Tule Lake and at Upper Klamath, but neither of these could be expected to replace Lower Klamath, and the duck hunters and nature lovers who had risen nationwide to protest the closing of the Klamath Strait were far from satisfied with the exchange. Besides, Tule Lake had problems of its own. What was left of the lake was serving both as a sump for irrigation runoff from the Klamath project and as a catch-basin for floodwaters from the Lost River, and its level fluctuated wildly from season to season. This played hob with the shoreline ecology, disrupted the growth cycles of the reeds and other marsh plants, and greatly lowered the lake's efficiency as a bird producer.

Too much water in Tule Lake: not enough in Lower Klamath. Hmmmmm. . . .

In November 1941, with the completion of the so-called Modoc Project, the problems were finally solved to nearly everyone's satisfaction. The key was a mile-long tunnel through Sheepy Ridge, plus eight giant pumps with an aggregate capacity large enough to handle the entire flow of the Colorado River in Grand Canyon. A system of dikes and canals and some land exchanges to block up the refuge lands in the deepest part of the old Lower Klamath Lake basin completed the job. Excess water from Tule Lake could now be pumped through Sheepy Ridge and used to recreate, at least partially, the waterfowl paradise that had been Lower Klamath. Judicious use of the pumps stabilized the water level in both lakes: The reeds returned, and so did the ducks. It was the last major land rearrange-

33

ment in the Klamath Basin, and it left things looking pretty much as they would during that damp spring fifteen years later when holdover worries about the drought managed to keep the farmers' attention to the point that few noticed how something far more dangerous and insidious than a drought was creeping up on them.

5

IN EARLY APRIL 1957, the clover fields—as they always were—were sprayed with DDT, mixed to a recommended dose of one pound per acre, to guard against a possible outbreak of root weevil.

Later that same month, the potato planting was begun. Great machines moved slowly across what had once been Tule Lake and Lower Klamath Lake, digging the holes and setting the seed potatoes at a rate of one per second, scraping the earth back over them as they moved on to repeat the operation eighteen inches farther on. In slightly under two weeks, 18,500 acres were covered this way, and the job was done. The total acreage came in about 10 percent under the 20,200 acres planted to potatoes in 1955, but no one worried much about the discrepancy. Because potatoes are demanding plants that quickly exhaust the soil they are planted in, it is normal to follow a four-year rotation pattern in which only one year out of the four is devoted to potatoes, with the others given over to the production of grains and of clover, and under these circumstances acreages in each crop are bound to vary a bit from year to year. Potato production, everyone knew, would be back up later.

In mid-May, the dry range was treated with 2,4–D to keep the brush down: one and a half pounds per acre for sagebrush, twice that for rabbit brush. Some of the irrigated pastures received a dose of the same chemical in what was already being recognized as a vain attempt to cut the dandelion population. "Each year," lamented Jendrzejewski in his "County Agents Report" column in the *Herald and News,* "there seem to be more." But there couldn't be any harm in trying, and maybe some day it would succeed. The dandelion spraying went on.

In late May and early June, the sprayers went into the fields with the first sprouting of the young potatoes and layed down a fine mist of Systox, one-half to three-quarters of a pound per acre, as an aphid preventive; shortly afterward, the clover fields were hit again (this time with toxaphene) to control lygus bug and seed weevil, and some time in June the city parks in Klamath Falls were given a quick jolt of DDT to rid them of mosquitoes. The children were out of school by now, and the days had grown long, hot, and lazy. Summer had begun.

In the county agent's office, meanwhile, a minor mystery had developed. For some years now, the county had operated a revolving-fund bait-mixing program, preparing poisoned grain—the hulled oats laced with strychnine recommended by Harold Schieferstein back in January to control the mice that were eating the fruit trees—and selling it to the farmers at cost, which currently meant about six and a half cents per pound. Most years this fund operated with a good healthy reserve, and there was plenty of bait to go around. This year, however, it was being strained to its limits—at a time when requests for its services should have been dropping down to next to nothing! The program's principal target was field mice, and field mice normally undergo a pronounced population drop in the spring. There is a simple mechanical reason for this. Field mice are extremely short-lived animals, with an average life span of barely eight and one-half months. They stop breeding over the winter in most years, and by the time breeding resumes in the spring natural attrition has cut their numbers considerably. Furthermore, since the spring breeding stock consists of animals that have survived the winter, the average age—and therefore, the average mortality rate—remains high for quite a while after breeding resumes, which makes for quite a pronounced lag between the resumption of breeding and recovery to normal population numbers. So there should have been a fall-off in the requests for poisoned grain. And that fall-off just wasn't occurring. The puzzled agents scratched their heads. *Why not? Where was all that bait going?*

In the end, they decided to blame it on the squirrels. Squirrels can be controlled with a hulled-oats-and-strychnine bait almost as easily as field mice can, and squirrels are much longer-lived animals; they don't show nearly as pronounced a spring population decline. This made them a far more reasonable culprit than the mice. Someone on the county agent's staff wrote an article explaining this connection and placed it in the *Herald and News* for May 26. It was a good explanation, and it held up for a full month. Then first haying time came along, and the phantom squirrel outbreak disintegrated.

Along with its offices in the post office building in downtown Klamath Falls, the Klamath County Extension Service operates an experiment station on Henley Road, a few miles south of town near the county airport; and it was here, in the low gray cinder block building that serves as experiment station headquarters, that I found Paul

Hatchett emptying desk drawers. Hatchett had been recommended to me as a source by Rodney Todd. A short, square-hewn, bespectacled man with a white crew cut, he was dressed in dark green work clothes, and a casual observer might have mistaken him for the janitor. He is, however, no janitor. He is a rodent-control specialist, and when I met him he was in the process of retiring from the position of Klamath County weed and rodent control supervisor, a post he had held since taking it over from Harold Schieferstein in 1964, fourteen years before.

In the spring of 1957, Hatchett was running a ranch in the Poe Valley, a few miles above Olene in the Lost River sub-basin, separated from the main part of the Klamath country by the enormous tilted bulk of fault-formed Stukel Mountain. We sat in a back room at the experiment station and talked about those days. "We had," Hatchett recalled, "about fifteen to sixteen hundred acres in hay and about twice that in permanent pasture. And we first noticed the damage—well, there was a buildup that went on for two or three years, but the stuff got rampant, the wildfire really hit along about first haying season. That's when we started poisoning. And then it just got worse and worse. . . ." All that long hot summer, while the farmers watched anxiously, the Klamath Basin proceeded to fill up with mice.

By early July the county agents had abandoned the squirrel hypothesis. They had to. "In the meadows, with all those tunnels and mounds," points out Walt Jendrzejewski, "why, you couldn't miss it." He wrote an article for the *Herald and News* identifying the real culprit and observing that "mice are very destructive when populations become excessive." That was an understatement. Damage was already so bad in the alfalfa that some growers had sprayed their fields with toxaphene in an attempt to eradicate the

Microtus population. The paper reported "excellent mouse control" by the use of this method. However, the mice didn't read this good news, and, in reality, the toxaphene failed to slow their population growth by any perceptible amount. "We cut them back some with the poisoning," admits Hatchett, "but we really weren't doing too much good." Destruction continued. Alfalfa fields slowly became riddled with holes; in the durham wheat, plants disappeared whole, cut off at ground level and consumed down to the last morsel. Mice scurried about everywhere, eating, tunneling, and breeding. By August they had moved into the clover fields, consuming the crowns and then moving right down the stalks to feed on the taproots. Some fields that had shown extremely high potential yields in July were not worth harvesting at all, while in others the clover seed obtained was barely enough to pay for the harvesting operation. But the worst news of all was reserved for September. That was when the potato harvest began, and it was discovered that mice had got to the potato fields as well.

Potatoes occupy a special place in the agricultural economy of the Klamath Basin, comparable to the place occupied by dairy products in Wisconsin or oranges in Florida—or potatoes themselves in Maine and Idaho. They are the basin's glamour crop. Though they cover only one-twentieth of the irrigated cropland, they account for more than one-sixth of the area's agricultural income. Klamath Basin potatoes are marketed as far away as the East Coast, and the "Klamath Gem"—a late-maturing potato of extremely large size and excellent flavor—is one of the best-known local varieties in the nation. The potato-growing area is concentrated in the central part of the basin, in the northernmost portion of the old bed of Tule Lake. Here soil and seasonal microclimate are exactly right, and here the small town of Merrill—once a lakeshore community, now stranded high and dry and surrounded by sixteen square miles of potatoes—holds an annual Potato Harvest Festival, a three-day bash in mid-October featuring parades, speeches, contests, exhibits, a harvest ball, and a football game which pits the Merrill High Huskies against the pride of Bly, or Bonanza, or Malin, or one of their other arch-rivals among small towns in the Klamath country.

One measure of the importance of potatoes to this area is their status in the Klamath County agent's office, where one of the assistant agents is traditionally given prime responsibility for potatoes and very little else. In 1957, that agent was Walt Jendrzejewski. There would be many times, over the next few months, when he wished that he wasn't. "They really hit the potatoes bad," he says today, shaking his head slowly as if the memory were still painful to him. "You'd dig into the hills and there'd just be a shell of the potato left . . . look out in a field, and, gee, it'd be nothing but holes and tunnels." Was such damage common? He shakes his head again. Field mice, he points out, are leaf-eaters, and do most of their damage to pasturage—grasses, hay, and alfalfa. "It's not normal to have much spud damage at all. Oh, you'll always have some on the lake bottoms, but they like meadows. But when they were spreading out all over, some of 'em found the potato fields, and I guess they found the eating was pretty good."

Pretty good is right. By mid-September, two weeks into the harvest, it was already being estimated that up to 5 percent of the crop would have to be culled due to mouse damage. That figure had to be constantly revised upward over the next few months, until by the time shipping of the crop had been completed in the spring of 1958 the overall cull figure had risen to almost 25 percent, with damage in certain fields running much higher than that. "They wouldn't be in a whole eighty-acre field, you know," says Jendrzejewski, "but in spots that amounted to maybe thirty to fifty percent of it." Other sources give figures as high as 80 percent, with destruction within those spots often almost total. In numerous places large potatoes would be discovered completely hollowed out, with mice nesting inside the thin shells that remained. In other places it would be found that the mice had gone through whole fields, nipping a small chunk out of each potato, as if testing for quality, and leaving the entire field fit only for livestock feed. The monetary loss amounted to several hundred thousand dollars; the psychological loss was incalculable.

The Potato Harvest Festival was left pretty much in a shambles. The body of it was there, but the heart had been eaten out by the mice, in much the same way that they had made culls of the finest potatoes. A petite Merrill brunette named Donna Carlson was crowned queen, to the accompaniment of a thirty-piece accordion band; Malin High School won the float contest in the Saturday morning parade, and the Huskies beat the Chiloquin Panthers that afternoon, 25 to 13. The local Ben Franklin store was voted first prize in the window-display competition. There was a barbecue, a carnival, a produce fair, and a "special honored guest"—a Mrs. Patricia Berry of Los Angeles, who had won a trip to the harvest festival on the old "Queen for a Day" radio program. But over it all hung a sense of forced gaiety, of dancing-on-the-edge-of-the-grave bravado. The Enemy was upon them: He was small and brown, with a short tail and beady little eyes. The Klamath County agents were still touting their oat/strychnine bait as "very effective," but over the line in California, Siskiyou County had distributed fourteen tons of it around Tule Lake without appreciably slowing the destruction, and in Modoc County, which had responsibility for the southeastern corner of the basin, they had abandoned strychnine altogether in favor of sodium fluoracetate, otherwise known as Compound 1080—quite possibly the most toxic substance known to man. Initial reports of success with this extremely dangerous chemical were glowing, despite the accidental deaths of several dogs—and not a few pheasants and other game birds that took the bait despite the fact that it had been color-coded to warn them away (mice are color-blind; birds are not). But in Modoc County, too, the mouse population continued to grow.

And now—as if the basin had not seen troubles enough—the growth rate itself, already high, seemed actually to begin increasing. It was a genuine population explosion, a mushroom cloud of mice, spreading through the basin in seemingly unstoppable waves. Proper mouse habitat had long since been overrun: Now they were overrunning everything else in sight. And for the people of the Klamath Basin as a whole—no longer just the farmers—it was the beginning of a nightmare that even today, twenty years later, most of them would still prefer not to talk about.

Near Henley in the central part of the basin, a rancher's wife turned over a couple of bales of hay after an October rain and was almost washed away by the flood of mice that poured out of them; she and her son killed more than one hundred by hitting them with shovels; the rest got away.

In the Langell Valley, on the upper part of the Lost River, a hundred fifty-ton haystack disappeared almost overnight.

Within Klamath Falls itself, flower beds were chewed up,

decorative shrubs were girdled, and even indoor potted plants were dug up and eaten. *Herald and News* editor Bill Jenkins reported finding a mouse nest in his yard roofed neatly with interlaced iris leaves taken from the bed next door.

At Fort Klamath, in the northernmost part of the basin, the cattle had to be moved onto winter feed a full month early because the pasture grasses had all disappeared down the gullets of the mice.

All over the basin, young fruit trees were dying—not just in isolated locations like those noted the winter before, but in great numbers, large-scale phalanxes of death. The mice ate the roots as well as the phloem layer under the bark, and toward spring whole orchards began falling over. . . .

"You can't imagine what it was like," says Dayton Hyde, shaking his head. Hyde runs an eight thousand-acre cattle ranch on the upper Williamson River, the stream that supplies most of the water for Upper Klamath Lake. The mice were there, too. "It was just a *mass of mice.* You couldn't keep ditchbanks, they tunneled them to pieces. You'd look out the door and you'd just see that movement, a seething mass of squealing bodies. You couldn't look at the ground without seeing mice. And they literally ate up everything. Haystacks would be just straw—fields looked like places where hogs had been rooting. They just literally demolished the ditchbanks, water was running everywhere. And if you walked out across a meadow, you'd find that the ground was spongy, like walking on a sphagnum bog. They also ate carcasses of larger animals, like sheep or cattle. In fact, anything you left down, like a glove or a pitchfork handle, just *went*."

Paul Hatchett's assessment, from his own ranch in the

Poe Valley, is similar. "It's just impossible to exaggerate how many mice there were around then," he says. "You'd find if you left a bale of straw on the ground three or four days, there'd be thirty or forty mice under each bale when you turned them over. And there was some human health hazard, everybody got their wells and groundwater tested for contamination and, oh, I suppose sixty or seventy percent of the farmers had to be tested for tularemia. . . . We had, I'd say, more than ten thousand acres of alfalfa plowed up and replanted on account of mouse damage. Some of that went into spuds, and some was replanted to alfalfa, but it was all on account of the mice. I've seen areas fifty-sixty feet square where I don't believe there was three square inches there wasn't a mound in."

Both Hyde and Hatchett are reporting events that have been filtered through twenty years of other memories, of course, but neither seems to be exaggerating the horror that was the Klamath Basin during the winter of 1957–58. Government accounts written at the time speak of mice constantly "scurrying over the surface of the ground," of mice that "frequently . . . can be observed crossing the highways at night," of mice that "riddled" ditch banks and drain banks with "tunnels and interconnected runways . . . from both sides." One official tally near Tule Lake counted over twenty-three thousand burrows in a single acre of alfalfa; another near Malin topped that with a grand total of 28,460, more than one burrow for every two square feet of ground surface. Up on the hills, as far as a mile from the damp meadows that the mice would normally never leave, damage to dry-range plants such as bitterbrush was so severe that game personnel began to be concerned about the ability of the local deer herd to find enough food to survive the winter. "There weren't too many in town," recalls Jendrzejewski, "but in the suburbs, my gosh, they were all over, and they even worked up on these hills, too, away from the flats. They girdled trees and actually killed them. It's amazing. . . . They were after the sapwood, you know, but they got the roots too. This isn't orchard country, but a lot of people have fruit trees around their houses, and when it became apparent what was happening people started watching. And a lot of trees fell—well, we've still got an awful lot of trees around, you know. But there were an awful lot killed." He shakes his head once more—there is a great deal of head shaking, even today, over those Klamath mice. "Boy," he sums up, simply and forcefully, "they were just all over the country!"

They were. By this time the basin looked like a war zone, the fields channeled and pockmarked as if by shell-fire, the ditchbanks riddled and leaking, the trees bare. Nearly everything green bore signs of massive damage. In certain areas, ranchers reported staring out of their windows at whole fields that were churning and surging randomly, as if the earth itself were in motion. And if you were very quiet you would swear that you could hear, faintly but unmistakably, a sound like the gnawing of millions of tiny teeth. . . .

6

BY MID-NOVEMBER 1957 the mice were page-one news in the *Oregonian,* the big metropolitan daily out of Portland, and the people of the Klamath Basin were beginning to discover that they were not completely alone in their misfortune. Though the Klamath country remained far and away the hardest-hit region, areas in many other parts of the West were reporting higher than normal mouse numbers as well. In Lake County, over the Gearhart Mountains to the east, ranchers in the Warner Valley were experiencing severe infestations; most of their stored grain was so badly damaged that it would no longer meet the standards of the Commodity Credit Corporation, the U.S. government agency responsible for administering price supports and food bank programs, and credit for it had been withdrawn. In southern Idaho, farmers reported circular bare patches as much as eight feet in diameter pock-marking their alfalfa fields. Central Washington's orchard districts, around Yakima and Wenatchee, had extensive girdling of young trees, and on Sauvies Island, in the Columbia River off Portland—an area not normally considered to be within the range of *Microtus montanus,* the animal doing all the damage—destruction reached appalling heights, with some alfalfa fields estimated to be as much as 75 percent destroyed. "This is the worst season we've ever had," reported U.S. Soil Conservation Service representative Tom Davis, with magnificent understatement, from the island. An *Oregonian* photographer went out with Davis and got a picture of a fox terrier dashing through a field of grain, the mice parting before him like the waves of the sea.

The plague was not without its quirkish moments. In the northern Idaho town of Moscow, a mouse crawled into the broadcast amplifier of radio station KRPL and brushed against the output transformer, frying itself and knocking the station off the air for more than two hours. Mice on the Gale Day farm near Amanda Park, Washington, moved inside in one sudden wave with the first fall rains and took over the house. "The walls, the shelves, the clothing hanging on the walls, and even my hat, had mice in them," reported Mr. Day. "Every bit of food had to be put in tin or glass. They even filled every plant dish with wheat stolen from the chicken house." In Klamath Falls itself, *Herald and News* editor Bill Jenkins waxed philosophical: "The current mouse invasion has some small good to it . . . . If one takes to stalking them with an adequate pellet gun it does much to sharpen the hunter's eye for the coming varmint season, not to mention giving him a little practice at looking through a pair of sights again. Spotting a wary mouse amidst the pine needles and cones is a tougher chore than one would think, and hitting them is not always easy. . . . Oh, well. It kills a few minutes here and there, anyway." But such levity was uncommon, and as time went on it became more uncommon still. Something big had clearly happened, something general enough to be felt in many places—and whatever it was, it was certainly playing hob with the normal population dynamics of the western field mouse. And even if no one could tell just what was causing it, actions were going to have to be taken against it. The mice were going to have to be destroyed. It was either them or us.

What *was* the cause? The editors of the *Oregonian* thought they knew, and they said so. In an editorial published on November 23, they laid the blame squarely on human activity, principally the "drastic reduction" in coyote and hawk populations. "The fecundity of mice is so great," they wrote, "that, if their natural enemies are reduced in number, the rodents overrun the country. Readers of the *Oregonian* may have noticed recently letters from upstate complaining of the shooting of hawks, apparently by trigger-happy hunters. . . . Another natural enemy of the field mouse is the coyote. From time to time complaints are heard from some ranchers that control measures against the wily prairie wolf are too severe, and that jackrabbits, field mice and other destroyers of crops and grasses are getting out of hand. A few years ago, a group of northwestern Colorado ranchers actually closed a forty thousand-acre grazing area to hunting and poisoning of coyotes. The ranchers explained they sought to regain a natural balance of wildlife; a drastic reduction of the coyote population had caused jackrabbits and field mice to increase tremendously. . . ."

The livestock interests responded immediately and vehemently. "You make the grave accusation that the cause of the rodent infestation in Klamath County was due mainly to the destruction of predatory animals and birds," wrote rancher W. H. Steiwer from the central Oregon town of Fossil. "As a matter of fact, I do not believe that there are any records to support such a theory, particularly in the case of the coyote." While admitting that "we do have a great number of mice," he insisted that "only three or four years ago we had just as many or maybe more. . . . While we do not have as many coyotes, we do have large numbers of bob cats. I am convinced that bob cats will eat mice, but am not too sure about the coyote. He might eat them when he cannot get anything else, such as a nice juicy lamb or young chicken or turkey . . . . The farmer who finds three or four young lambs at one time killed by coyotes, or one or two of his best laying hens, would be hard to convince that we should let the coyotes live in order to protect him from a future rodent infestation. I rather think he would say, 'I'll take the mice.' "

The Steiwer letter was the only one that saw print in the *Oregonian*, but it was almost certainly not the only one received, and there were plenty of comments made at cattlemen's meetings and in the privacy of ranch headquarters. Even some of the newspaper's own staff was clearly upset by its stand. "Since a normal coyote population is about one per square mile," pointed out one staff writer several years later, apparently after brooding about it for some time, "and the field mice were estimated at six to ten million per square mile, this hardly makes good arithmetic, much less sense." Clearly, the *Oregonian* had touched a sensitive point. It had dared to assail one of the great sacred cows of the West, and it was going to be told about it. It was of infinitesimally small consequence that the paper might be right.

A *predator* is any animal that kills another in order to eat it. A mountain lion ambushing a deer, a wolf pack running down a fleeing moose, and a bobcat chasing a squirrel, are all predators; so is a dragonfly eating mosquitoes, a spider waiting in its web for a passing fly, a house cat creeping up on a sparrow, or a human pushing buttons in an automated slaughterhouse. Swallows, shrews, and insectivorous bats are predators, as are those animals—skunks, squirrels, robins, pigs, and many others—that only kill for food part of the time. There are even predatory plants—the Venus's-flytrap is an obvious example—and some ecologists go so far as to class animals that eat plants as predators, with the plants as prey.

When we speak of *predator control,* however, we are using the word *predator* in a much more specific and limited sense. Though there have always been a few individuals who consider every nondomesticated animal an enemy that must be eradicated at all cost, the great majority of farmers and ranchers see no harm in the smaller predators, and most even think of them as at least mildly beneficial. It is not difficult to grasp the fact that a barn owl's diet—for example—is mostly small rodents, and it doesn't take a degree in ecology to understand that this diet must be doing more good than harm. Most farmers have always spoken with open disdain of the hunters who bang away at hawks and owls for target practice and then leave the carcasses hanging on their fences. And when mice are particularly numerous—as they are, because of the field-mouse cycle, every three or four years—the newspapers in farming country regularly carry letters to the editor suggesting that the shooting of birds of prey be outlawed, at least until after the mice are gone. What concerns livestock owners is not predators in general, but the larger ones, the ones capable of taking a sheep or a cow—the bears, the mountain lions, the eagles, the bobcats, and the wolves. And the coyotes. Especially the coyotes.

In the demonology of the West, the coyote has always occupied a special place right in the center of the innermost ring of hell. He is the "coyote pest," the "coyote scourge," the "coyote menace." In livestock owners—and especially in sheepmen—he inspires a flaming hatred that passes all bounds of rationality. They paste bumper stickers on their cars and pickups that read, "Eat American Lamb—20,000 Coyotes Can't Be Wrong." They carry rifles in their vehicles constantly on the off chance that they might get a shot at him. They hunt him, they trap him, they poison him, they chase him with airplanes, they lace their ranches and the public grazing lands with "coyote-getters"—a device like a zip gun that is buried in the ground and set to shoot cyanide-tipped darts into the mouth of any animal curious enough to tug on its baited wick—and in between times they write letters to the newspapers and to the Bureau of Land Management and to their congressmen demanding higher levels of funding so that there can be more hunting, and more trapping, and more chasing, and more coyote-getting. Otherwise decent men, men fond of their wives and children and tenderly solicitous toward dogs, horses, and sheep, have been known to gleefully wire captured coyotes' mouths shut and turn them loose to starve, or to saw off their lower jaws with hacksaws, or to bind gunnysacking around them, douse it with kerosene, and set it afire, or to purposely leave them in traps for up to a week after they have been captured, waiting for their wounds to putrefy and their bodies to dehydrate, making sure that they suffer before they die. Robert E. "Pud" Long of Odessa, Oregon, the top government trapper in the United States during the 1950s, described by neighbors and friends as kind and gentle, used to kill coyotes he trapped by hitting them with the blade of a shovel until they were too stunned to move and then stomping them to death with his high-heeled cowboy boots. He averaged nine kills a day this way, and once set a record of forty-two. (Strangely enough, this practice was reported admiringly in a January 26, 1947, article in the *Oregonian* by writer—later U.S. Senator—Richard L. Neuberger, in whose name the Oregon

Environmental Council today gives an annual award.) And while many stockmen deplore this type of needless cruelty as much as anyone, the killing itself is cheerfully condoned by the vast majority of them, who insist that without it they would long ago have been put out of business because of livestock losses.

What sort of animal is it that inspires such hatred? Who is the beast that eats lambs alive, that drives sheepmen to ruin, that appears out of the night and kills whole herds of animals, "just," as one 1912 writer put it, "for the lust of killing"—and of whom two out of three stories such as those mentioned earlier in this sentence will be untrue? He is usually thought of as just a small wolf—the names "brush wolf" and "prairie wolf" appear often in the literature—but those who expect the shining eyes, the hulking shoulders, and the slavering jaws of the wolf of folklore will be vastly disappointed by their first sight of him. The typical coyote is small (twenty to thirty pounds), dun-colored, and shy, with a slender, rather weak-looking jaw and a bushy tail that amounts to about a third of his four-foot length. In *Roughing It*, Mark Twain called him a "long, slim, sick and sorry-looking skeleton, with a gray wolfskin stretched over it," which is not far from the truth, although Twain's further characterization of him—"even the fleas would desert him for a velocipede"—seems a bit harsh. Surviving for ten or twelve years in the wild, usually mating for life, he lives in small packs on his denning grounds but rarely hunts in groups of more than two or three. His pups, anywhere from two to ten to a litter, are born blind and helpless but mature rapidly: Weaning begins by the third week and sexual maturity may come as early as the ninth month, although pups born to a yearling mother are apt to be less strong than others. His most outstanding characteristic is his voice. Wide-ranging, extremely musical, with good pitch control and excellent tone production, it is perhaps the most beautiful of mammalian vocalizations, not even excepting man's. As a trained musician with a specialty in voice, I am constantly amazed at the incredible vocal gymnastics performed by these animals, and at the quality of the sound that seems to be produced so effortlessly. So, evidently, were the nineteenth-century scientists who first studied him. They called him *Canis latrans*—the barking dog, the songdog.

Of course, all the vocal ability in the world wouldn't redeem the coyote if he were the menace to sheep and cattle that the stockmen say he is. Is he? Here I must admit to a bias: I have been fond of coyotes, which I consider to be improved versions of the domestic dog, ever since I saw my first one in the Portland zoo at the age of seven. However, with this bias in mind, I recently spent several months searching through statistical data banks, scholarly journals, and scientific libraries for evidence that might prove me wrong, and I will still say that the vast preponderance of evidence is on the side of the coyote. Study after study has shown that the animal is primarily a small-mammals consumer, and that anything much larger than a rabbit is a very rare item in his diet— except when it comes to him as carrion. Charles C. Sperry examined nearly fifteen thousand coyote stomachs from seventeen states in the course of a five-year study for the U.S. Fish and Wildlife Service in the early 1930s. His results, published in 1941, show rabbits as the top food item, accounting for one-third of the total, with carrion coming next at one-fourth. Rodents were third—at 18 percent—while domestic livestock ran a poor fourth, amounting to only 13.5 percent of dietary items even though the stomachs examined all came from animals taken by predator-control officers who thus could be expected to be biased toward the heavier livestock users.

A similar study done by Adolph Murie in Yellowstone National Park at about the same time, involving the analysis of 5,086 coyote droppings, put field mice in first place at 34 percent of the diet, followed by gophers at 22 percent. Large mammals as a group, including elk, deer, bighorn sheep, antelope, bear, and domestic livestock, came to 19 percent of the total—"mostly," Murie wrote, "as carrion"—while the remaining 25 percent was made up of birds, grasshoppers, vegetable matter, a house cat or two, and an astonishing horde of miscellaneous nonfood items eaten by the animal for God only knows what reason: paint-soaked rags, shoelaces, tinfoil, banana peels, gloves, mud, rope, and a seven-inch swatch of somebody's living room curtain. Most recent work—Hope Ryden's two-year follow-up of Murie's study in Yellowstone in the early 1970s, Frederick Wagner and Charles Stoddart's eight-year, twelve hundred-square-mile examination of the relationship between coyotes and jackrabbits in Utah in the late 1960s—has reached the same conclusions that Murie and Sperry did. Not just one, but *two* blue-ribbon presidential commissions (the Leopold commission in 1964, the Cain commission in 1971) have stated quite explicitly that predators in general, and coyotes in particular, pose no serious threat to livestock owners, and that predator control as currently practiced does more harm than good. ". . . No coyote food-habits study," reads the report of the Cain commission, in part, "has ever shown livestock to be a major part of the diet. . . livestock predation, over an entire population, is an infrequent event."

On the other hand, studies purporting to prove that the coyote is a serious hazard to livestock almost invariably turn out, on close examination, to have many more holes in them than the coyotes' teeth are supposed to have put in the livestock. One group of researchers, for example, took some captive, pen-reared coyotes, starved them for periods ranging from several hours to several days, and released them into a small pen containing sheep. In twenty out of thirty-eight trials, or very slightly more than half the time, one or more sheep were killed, whereupon the authors dutifully published their results in the *Journal of Wildlife Management* as proof, among other things, that "most wild adult coyotes are able to kill sheep." In a later series of tests, published in the same article, well-fed coyotes were turned in among the sheep, and a similar ratio of kills resulted as when the coyotes had been starved, which would seem to prove the authors' contention that coyotes will kill even when not hungry. Unfortunately for their case, however, the authors seem to have used the same coyotes for the second set of tests as for the first set—proving only that when an animal is taught to kill in a particular situation, he will kill again when the situation is duplicated.

Another group took two areas in south Texas, practiced intensive predator control on one but left the other alone, noted that small-mammal populations remained the same on the two areas but that doe/fawn ratios improved dramatically on the "treated" area, and concluded that coyotes have more effect on deer populations than on small-mammal populations. This conclusion looks valid until you read further and discover several things. First, the "treated" and "untreated" areas were less than a square mile each (compared to a normal coyote hunting range of approximately eighty square miles) and were less than five miles apart, close enough so that predator control on one area would influence predator numbers on the other. Second, the study was only two years long, not nearly long enough to take into account the effects of natural small-mammal population cycles (as discussed earlier). Third, the entire study area "provided"—in the authors' own words—"unsuitable habitat" for jackrabbits, which are a mainstay of coyote diet elsewhere. Finally, the "untreated" pasture was right next to ranch head-quarters and barely a mile from an area where heavy exploratory oil drilling was taking place, a combination of factors which, the authors admit, reduced the deer population by approximately 80 percent during the study period and thereby (although they don't admit this) knocked all of their carefully prepared results into a cocked hat.

But if the coyote is not a significant hazard to livestock —and no serious study indicates that he is—whence then this terrible, irrational hatred? Part of it is based on fact: Individual coyotes do kill individual sheep, often horribly. There have been plenty of eyewitnesses to these killings and some of the eyewitnesses have had cameras with them. Few of their accounts bear much resemblance to folkloric tales of predators dispatching their prey quickly and cleanly. Predation, as François Leydet has pointed out in his excellent book *The Coyote: Defiant Songdog of the West,* is not an act of mercy but one of pragmatism: To the predator, the prey is only a food object, and thus needs only to be incapacitated—not killed—before feeding can begin. This leads to many situations where lambs and calves are eaten alive, revolting the human observers who do not share the value system of either the coyote or its victim.

Another cause of the hatred is honest misinterpretation. Coyotes, being carrion eaters, are often seen feeding on the carcasses of sheep dead from other causes—diseases, de-hydration, a fall—and blamed for killing them. Some authorities believe that as many as 75 percent of reported "kills" may fall into this category. It is also possible to observe coyotes chasing stock for reasons that have nothing to do with attempting to kill the stock, and to leap to the conclusion that a kill is imminent. I have done this myself. Once, while driving over Table Mountain in the southern Oregon Cascades, I came suddenly upon a solitary coyote chasing three adult cows and said to my wife, without hesitation, "Look at that coyote chasing those cows!" It was only later that it occurred to me that a lone coyote would have an extremely hard time killing anything as large as a single cow, let alone three of them, and that what the coyote was probably actually engaged in was "mousing"—running the cattle across the pasture and pouncing on the mice that leaped out of the way of their hooves. This is a coyote hunting technique that has often been observed, with elk taking the place of cows, in places such as Yellowstone National Park.

Still a third cause for the hatred is scapegoating, blaming the coyote for the demise of an industry that is dying for other reasons, notably the invention of synthetic fabrics and the consequent drastic decline in the demand for wool. The number of sheep raised in this nation has been dropping precipitously for forty years now, from a high of around 50 million animals in the late 1930s to 27 million in 1956, 20 million in 1970, and 16.5 million in 1974, the lowest point since the Civil War. These numbers would probably be even lower if it weren't for government price supports, which have been in effect since passage of the National Wool Act in 1954. Blaming the coyote for this industrial morbidity is foolish, especially when you consider that coyote numbers have also declined—with no perceptible effect on sheep trends. But such blame continues to be placed.

However, it is the fourth cause of the hatred that is probably the largest of all, and this fourth cause is revealing. It is, simply, the shamefully infuriating fact that the coyote is so adept at making those who try to get rid of him look like fools.

Set out poisoned animal carcasses and he learns to avoid carrion—and then, because he is hungry, he raids your sheep herd twice as often. Set out traps and he learns to detect them, digging them up and turning them over to render them relatively harmless and as often as not defecating on them, apparently for no other purpose than to tell you what he thinks of you. Hunt him by air and he learns to hide at the approach of an airplane, sometimes by diving in among your livestock herds where you can't get a shot at him without risking your own expensive animals. Outsmart him, outtrick him, manage actually to kill him—and his remaining relatives will respond by upping their birthrates by a factor of at least four times. One pair of researchers, Guy Connolly and William Longhurst at the University of California, estimate that if you were able to eliminate 75 percent of the coyote population each year it would still take a full fifty years to render the animal extinct. Some of the more literate ranchers like to refer to him as Don Coyote, and the name has a certain rightness to it: Certainly he has the ability to make the rest of us feel a bit like Sancho Panza.

Not all ranchers hate coyotes, of course. One who doesn't is my friend Ogden Kellogg, who has run a three hundred-acre cattle farm on southern Oregon's Sardine Creek, near the small Rogue River community of Gold Hill, for more than twenty years. Sleeping on the lawn beside Og's big white ranch house not long ago, I was awakened at two o'clock in the morning by one of the loveliest coyote choruses I have ever heard. I told Og about it later, and he just smiled. "I kinda like 'em," he said. "Don't tell my neighbors."

The Carson Factor

Another friend of the coyote is Dayton Hyde, the Williamson River rancher whose graphic descriptions of the mouse hordes highlighted our discussion of conditions in the Klamath Basin as of the fall of 1957. "It's been my experience that predator losses start with predator control," he told me over a cup of tea in a small lakeside cabin on his ranch one harsh February day. "You keep coyotes with a balanced population and food supply, and you won't have any problem. We've actually been trying supplementary feeding of predators in January and February, leaving carcasses out where they can get at them. We don't have any predator loss, and the coyotes are among our cattle all the time. And we no longer do anything to deter mouse populations, because we don't have to. The trouble isn't coyotes. The trouble is—" He leaned back in his swivel chair and grinned. "The trouble is, we're raising smarter and smarter coyotes and dumber and dumber people."

This, unfortunately, represents a kind of enlightened attitude that one does not run into very often. And if it is rare today, it was even rarer in 1945, as World War II was just ending—and what we will now begin calling the Great Coyote War was just getting under way.

In calling what happened to the coyote in the decade following World War II a "war," I am breaking no new ground: That is precisely what the newspapers of the day called it. AIR WAR HITS POOR COYOTE, trumpeted one headline. COYOTE WAR GETS NOTICE, bleated another. WAR ON COYOTES, proclaimed a third. It was not just an effort to reduce the population, but a blatant, all-out attempt to get rid of it completely. "Eradication of the coyote," wrote one reporter after interviewing the troops, "should not be considered on a cost basis but rather on the saving to the livestock industry and replenishing the deer supply." "Extermination, rather than mere control, is the aim," stated another. In the last section, we took a look at some of the general attitudes that could lead to a genocidal attempt of this magnitude. What we must do now is examine the situation in the mid-1940s that caused a sudden escalation of the war, and get a good look at the new weapons that very nearly, despite the great odds, gave it the power to succeed.

In a sense, of course, there was nothing new at all about this "war on coyotes." Predator control did not spring full bodied out of the earth at the close of World War II; it had been around a long, long time before. Humans have been attempting to eradicate their competitors for meat ever since the first Cro-Magnon domesticated the first cow. Analogies equating predators with evil and predator control with acts of virtue are a recurrent theme in the Bible (the Good Shepherd), in Aesop's Fables (the Boy who Cried Wolf, the Wolf in Sheep's Clothing), in Roman mythology (the word *predator* itself comes down to us from the Latin *praedator,* which means "plunderer"), and, indeed, in most of world literature. These attitudes came to America with the first settlers. In his journal entry describing the landing at Plymouth Rock, William Bradford tells of the Pilgrims' fears concerning this "hideous and desolate wilderness, full of wild beasts and wild men"; the Colony of Massachusetts had a thirty-shilling bounty on wolves as early as 1648, and the other colonies—and the states—soon followed suit. The bounty system spread across the nation in a slow wave with the advancing settlers, reaching Oregon in 1848, where the first territorial government grew out of a wolf-control meeting at what is now Champoeg State Park. The first modern bounty law was passed in the state in 1909, and during the next four years nearly ninety-two thousand coyotes were slaughtered here for the $1.50 per pelt payment. Still, it did not seem to be enough: As of 1912, the coyote remained "the worst pest the state has to put up with," costing the livestock industry an estimated $25 thousand annually. A perusal of some of the more sensational newspaper articles dealing with the problem makes it difficult to imagine how the sheepmen thought the bounty system was doing them any good at all. "Frequently a band of from 20 to 100 sheep will get separated from the main herd," wrote one reporter of sheepherding in eastern Oregon. "If this band of sheep is not brought back to the main herd before night there is absolutely no chance of their surviving until morning." Coyotes were "known" to kill up to seventy-five sheep in one night: "He cuts their throats for the lust of killing. He seems to have a mania for murder." Heretofore confined to the dry eastern portion of the state, the animal was extending his range, "spreading his career over the Cascades into the Willamette Valley, where he will be an infinitely more serious pest than he is at present." This range extension—which was real—was probably caused by the ranchers' own predator-control activities: By exterminating the wolf, they had created a predatory vacuum in the western part of the state, and the coyote was moving in to fill it. This, of course, was not understood, nor was the Willamette Valley sheep industry's real problem—that the sheep owners refused to hire herders, preferring to graze their sheep in small solitary bands that were supposed to take care of themselves. The coyote killing went on, and cries went forth for it to be increased.

In 1914 the federal government got into the act, as Congress passed a law authorizing the expenditure of $125 thousand in salaries for predator-control officers on the western rangelands. This sum leaped upward over the years, until by the 1940s an amount equivalent to that original authorization was being spent on trappers' salaries and supplies in the state of Oregon alone. It did little apparent good; the reported depredations of the "coyote scourge" continued to increase right along with the money being spent to stop them. The *Oregonian* for January 16, 1944, for example, speaks of a "tremendous increase in the prairie wolf family in recent years." The reporter quotes Paul T. Quick, regional director of predator control for Oregon, Washington, California, Nevada, Idaho, and Montana: "An increase in coyote population invariably results in increased destruction of livestock and poultry." Borne forward by statements such as this, with the weight of Quick's authority behind it, even city folk could begin to feel some of the desperation of sheepmen as they battled the "four-footed plague" that was ravaging their industry. Clearly, something would have to be done. A "war on coyotes" seemed both inevitable and honorable.

Was all this hue and cry justified? Was there, despite all the killing—a million and a half coyotes dead at the hands of federal agents alone, never mind the bounty hunters and the itchy-fingered ranchers, in the three decades between 1915 and 1945—a real increase in the number of animals roaming the West during that first half of the 1940s? Most ranchers and predator-control agents claimed that there was, giving as the probable cause the lack of available manpower and ammunition caused by World War II, but the kill figures published by the old Predator and Rodent Control branch (PARC) of the U.S. Fish and Wildlife Service fail to back the claim up: They remain almost constant through the war. However, quibbling over these figures probably misses the point, because no matter what was happening to the coyote population, the number of coyotes per sheep in Oregon was clearly on the rise. Coyote numbers did not have to increase to bring this about. Sheep numbers were going down.

The sheep industry was in a tailspin. Statewide, the number of sheep had declined drastically, dropping approximately 60 percent in the twelve years between 1935 and 1947. Part of this was due to the popularization of rayon and nylon, part to a war-caused depletion of the manpower pool from which the sheep owners had been drawing their herders, and part to a definite swing in the pendulum of national taste away from mutton and toward beef. But it was more convenient to blame it on the coyote, and the sheep industry, as a body, did just that. Besides, the coyote was something that it seemed the industry might be able to do something about. The last decade had seen a major revolution in the coyote killers' weaponry, and for the first time in the eons-long contest between human and predator the humans looked as though they just might be able to win.

Three weapons were central to this new arsenal. First of these was the "coyote-getter," already mentioned. Invented by one Fred Marlman in 1939, the "getter"—also known as the "cyanide gun" or the "M–44"—consists of a steel tube about a foot long, a spring-loaded plunger, a cyanide-tipped dart, and a trigger mechanism fastened to a bait wick that is scented with carrion. The device is buried in the ground with just the wick showing, aimed so that an animal pulling curiously at the wick with his teeth will get the cyanide dart in the roof of his mouth. Usually the animal that pulls the wick is a coyote, but sometimes it is something else—a dog, a bobcat, a skunk, a sheep, even a human. Tens of thousands of these contraptions were spread across the western range during and immediately after World War II.

The second tool was a substance called sodium mono-fluoracetate, better known as Compound 1080. Developed by army poisons researchers during the war, 1080 became available to PARC agents and western stockmen about 1944. It is cheap (twenty-eight cents will buy enough to kill eleven thousand coyotes), easy to use (one simply injects it into an animal carcass, which then becomes known as a "1080 bait station" and is left someplace out on the range), and extremely effective (.0014 grams will kill an adult coyote). It is stable: A dose of it left out in the open for twenty years is every bit as toxic as a dose placed out yesterday, and an animal that feeds on an animal that feeds on an animal that has fed at a 1080 bait station is nearly as likely to die as the creature that originally ingested the lethal dose. It is odorless, colorless, tasteless, and virtually undetectable. There is no known antidote.

1080 was extensively used over much of the western range, but it was never very popular in Oregon ("1080? Oh, we stayed away from that stuff, we were afraid of it," Walt Jendrzejewski told me in no uncertain terms). And for this reason it is the third new tool—aerial hunting—that is most important to our tale of the mice.

The term "aerial hunting" refers to the technique of chasing down coyotes with a light plane and shooting them as they run. Originated in South Dakota around 1942, apparently by a group called the Castle Rock Coyote Extermination Association, aerial hunting quickly spread to the other western states. It was introduced into Oregon toward the close of World War II by a mechanic named Floyd Capp who worked for an auto dealer in Burns, out in the middle of the eastern Oregon rangeland. Beginning in January 1944, Capp and his friend Oscar Davis, an expert pilot and wartime flight instructor, flew out into the desert around Burns almost daily looking for coyotes. "A lot of sport," they told the *Oregonian.* Another Burns resident, W. M. Bennett, provided details on that "sport": "The gunner has to sit low in the plane," he told newsmen, "with only a small opening through which he can aim. His target field is only about ten square yards in extent. And when the coyote dodges out of that area the hunter doesn't even get a shot. . . ." It was also sport for the pilot, who had to fly low to the ground, turning and twisting with the coyote's course and attempting to get him to run in a straight line so that the gunner, firing out the left-hand window, could get a good shot at him. ("Aim for the tip of the tail," one reporter was told, "so that the plane's speed relative to the coyote's will give you a good pattern of shot.") In fact, just about the only individual it wasn't sport for was the coyote. Capp and Davis killed 278 coyotes in thirty-five days during the first two months of 1944, and the local ranchers, previously skeptical, were convinced. Soon the duo was hiring out its services at $12.50 per hour. The air war on coyotes had begun.

Other aerial extermination teams sprang up rapidly around the state. By March 1945 the Oregon Game Commission had three crews of its own hunting coyotes from the air—"primarily," according to game commission spokesman A. V. Meyers, "to protect game animals and birds," but also as "relief for farmers and stockmen." PARC agents quickly got into the act, too. "We're having excellent success using airplanes to rid the Oregon countryside of coyotes," crowed Harold Dobyns, chief coyote-control agent for the Oregon branch of PARC, early in 1947. And indeed they were. For the first time since intensive coyote-control efforts had begun, the number of coyotes in the state was actually going down. It is next to impossible to pin numbers on that decrease, but we can get an idea of the general trend by examining PARC kill reports for Oregon in the years following World War II, which dropped steadily from 10,786 coyotes killed in the fiscal year ending June 30, 1945, to a low of 3,838 in the fiscal year ending June 30, 1957. Since the overall effort actually went up slightly during that period—forty-one agents in the field at the end versus forty at the beginning—it is safe to assume that the drop in kill figures represents a real drop in population.

How much of a drop? You can do some mathematics based on known coyote population models, such as that developed by Frederick F. Knowlton and published in the April 1972 *Journal of Wildlife Management,* and come up with an approximation. Natural, uncontrolled coyote populations have an adult mortality rate of 40 percent, which means that four out of every ten adult animals will die during any given year. Mortality among the pups is slightly smaller: It works out to about 33 percent. In a population harried by intensive control efforts, about 70 percent of all adult females—that is, about 35 percent of the adult population, given the known coyote sex ratio of fifty-fifty—will give birth to a litter of six or more pups each year. Plug all these figures—and the annual PARC kill figures—into the equations used by population biologists to calculate changes in population size, press a few buttons on a pocket calculator, and you will find that Oregon's coyote numbers decreased by a factor of about 60 percent during those critical ten years between 1945 and 1955. The Klamath Basin, of course, got its share of that decrease—a share that works out, if the standard "natural" population density of one adult coyote per square mile of habitat is assumed, to about six hundred animals. If the coyote really was a menace to sheep, and therefore best got rid of, humanity certainly was doing a good job of it. No one appears to have been too concerned about what else they might have been doing.

By the mid-1950s, certain signs of predator-prey imbalance had begun to show up in the Klamath Basin. Deer herds had increased by nearly 40 percent since 1950, despite a 65 percent increase in the number bagged each hunting season, and were beginning to cause a significant amount of damage to farm crops; one rancher, Edwin J. Casebeer, estimated that deer depredations were costing him two thousand dollars annually. This increase showed no sign of peaking out. Pocket gophers were also causing problems—the county agents suggested poisoning them with carrot strips laced with warfarin—and porcupines had become a serious menace, especially in the forests where they had apparently become quite fond of the taste of young conifers. The Weyerhaeuser Company, which owned a great deal of forest land in Klamath County, claimed to be losing one out of every fifteen seedlings it set out to the big prickly rodents, and they organized hunting parties to seek them out and club them to death wherever they could be found; Klamath County authorities offered a fifty-cent-per-nose bounty, which resulted in the demise of fifteen thousand porcupines in less than six months and supplied spending money for a large percentage of the county's school-aged children, but failed to make an appreciable dent in the depredations. There was an elk overpopulation problem in some other parts of the state, notably the Blue Mountain region in the northeastern corner, and it was threatening to show up locally as well. And there were the mice. We musn't forget about the mice.

7

NO ONE IN THE Klamath Basin was forgetting about the mice. No one had much chance to. The mice were everywhere, eating, tunnelling, breeding, and generally creating havoc. "Conservative estimates," wrote County Agent Henderson in a special report to the U.S. Public Health Service, filed during the second week of November 1957, were "millions of mice . . . 400,000 infested acres in Klamath County alone. The damage is still continuing in the clover stands with the mice tunnelling in the soil, eating crowns and roots. The killing of plants is perhaps even worse in the alfalfa fields. . . . This is even before the continued action of the mice during the winter. . . . Grain fields . . . are being riddled, some with areas an acre in size or greater as dead patches. Many ditches and drains throughout the basin will probably not hold water next season. It is hard to find a ditch that does not have the banks honey-combed with holes. . . ." Half a million dollars had already been spent for control with "no appreciable diminution in numbers." And things were going to get worse before they got better.

Breeding season should have been over long before, but the mice didn't know this. An extraordinarily large number of individuals captured in the last few weeks had turned out to be pregnant females. It was going to be a long winter.

Henderson's report was important not only for its graphic descriptions, but for what it signified. A new phase had begun in the Great Mouse War of Klamath County—a phase which, depending upon how you looked at it, could either be considered cautiously hopeful or the most desperately alarming of all. The huge wheels of the bureaucracy were beginning to grind: Finally, after months of receiving and filing reports such as Henderson's, it appeared as though the federal and state governments might be moving to come to the aid of the mouse-beleaguered Klamath ranchers and townspeople.

We must not be too hard on the bureaucrats, who after all were only doing what they had been asked to do. In fact, their actions during that Winter of the Mice were in many ways models for what government behavior ought to be: They came into the basin only when asked, they confined themselves largely to their areas of expertise, and they worked with admirable efficiency and with frank and open acknowledgment of their own shortcomings. It was not their fault that the Oregon Meadow Mouse Irruption of 1957, as it was now being called officially, did not follow the book on how mouse population dynamics were supposed to behave—and it was certainly not their fault that the Oregon gubernatorial election was coming up, and that the mice were, in an otherwise dull period, a perfect piece of raw material for shaping into a series of political footballs.

The state of California leaped into the fray quickly;

state agencies there had become active in fighting the mice in their part of the Klamath Basin as early as six weeks before Henderson filed his report and the federal government began to admit, rather belatedly, that something unusual was going on. In late September, the California State Division of Agriculture had sent an employee named Rollo Talbert to the Tulelake area for a two-day investigation that culminated in a recommendation of massive amounts of 1080—the coyote wonder-killer that had come out of World War II. Modoc County began aerial distribution of 1080-soaked grain on October 1, and by mid-November more than thirty tons of this hellish substance had been spread over about six thousand acres of irrigated farmland in the southeastern sector of the Klamath Basin. On the farms, children were kept indoors, and hunters were advised to keep their dogs out of the fields.

Siskiyou County was more cautious. As soon as word came down that 1080 would be permitted, and even encouraged, Deputy Agricultural Commissioner Bill Huse—worried about the environmental effects of wholesale use of this extremely persistent and toxic chemical —bucked a recommendation up to the Siskiyou County Board of Supervisors in Yreka that they ban the use of it. They did. Siskiyou County would get by with lesser materials.

In Oregon, meanwhile, it had begun to look as though they might have to get by with nothing at all. Despite pleas from the farmers and ranchers, and from the county government, the state department of agriculture had taken the rather curious position that the mice were not within their jurisdiction. The mice might be overrunning everything in sight; they might be decimating the potato crop, quartering the clover seed harvest, causing thousands of acres of alfalfa to be plowed up; they might be infesting half a million acres of farm land and doing an estimated $5 million worth of damage; but it was not a Department of Agriculture problem. It was a health department problem. In Oregon, nothing official would be done about the mice until and unless the State Board of Health decreed that there was a significant human health hazard in the mouse area.

Fortunately, the Board of Health was equal to the challenge, which was perhaps not so strange as it seems. Mice are, after all, a bona fide health concern: They carry a long list of human diseases, including typhoid, tularemia, rabies, rat bite fever, and bubonic plague, and the greater the concentration of mice, the higher the incidence of these diseases. So if it was unusual for the department of agriculture to wash its hands of the problems raised by a multimillion dollar crop loss, it was not so unusual to expect the health department to be interested. Health agencies on both state and federal levels were, in fact, watching the situation quite closely, on several levels; and in early November, they began to move.

Early in the week of November 10, Dr. William R. Jellison of the U.S. Public Health Service's Rocky Mountain Disease Laboratory in Hamilton, Montana, arrived in the Klamath Basin, and with the assistance of local health officials and personnel from the Klamath County Agent's office, captured two hundred mice, put them in Mason jars, and took them back to Hamilton for study. Fifty were immediately killed and autopsied; the remainder were kept in their jars and watched closely for signs of disease.

On Friday, November 15, Dr. Jellison returned to Oregon, to a special meeting of state and federal public health officials in Portland called specifically to discuss the mouse problem, which the newspapers were now beginning to call a "plague." Three of those fifty autopsied mice, Jellison reported, had turned out to be carrying disease, although the exact nature of the disease had not yet been determined because the pathology lab had not yet had time to culture the disease organisms. The scope of the problem as a whole, however, was clear enough. It was, Jellison told the assembled officials, the largest field mouse infestation ever reported in the United States and quite probably the third largest in the entire world; the only ones that had definitely exceeded it were two that had hit Czarist Russia in the closing decades of the nineteenth century. Unenviable as its method for getting there was, Klamath County was now definitely on the world map.

The lack of disease was a mixed blessing. It meant that there was as of yet no appreciable health hazard to humans, but it also meant that there was no appreciable health hazard to the mice. Disease is one of the principal natural control agencies that keeps mouse populations in check. An animal with a hyperactive birth rate needs a high death rate as well if its numbers are to remain relatively constant. Epidemic disease is one of the best means of insuring that high death rate; without it, you have to depend on other, less trustworthy means, or suffer a population explosion. The Klamath Basin was doing the latter. So the absence of disease, while desirable from a health standpoint, was undesirable from all other standpoints. Despite the hazard, the search for what the *Oregonian* called an "antimouse disease" would have to be continued.

On November 18, one of Henderson's assistants, J. D. Vertrees, spent all day walking through the fields of the Klamath County portion of the mouse area looking for dead mice. Living ones scampered everywhere; dead ones seemed totally nonexistent. He failed to find a single one.

On November 20, there was a faint ray of hope. One of the one hundred fifty mice that Dr. William Jellison was keeping in Mason jars in Montana died; an immediate autopsy was performed, and the cause of death was determined to be tularemia, one of the major epidemic diseases of field mice, and one that has been linked at other times to rapidly declining mouse populations. Jellison immediately telephoned the news to the Oregon State Health Department, which passed the news on to Klamath County Health Officer Dr. S. M. Kerron. The newspapers announced it the next day. But it was only one case, and it was in the lab—not in the field. Even when four more mice died over the next ten days, it seemed a bit premature for celebration. There was always the chance that the tularemia could have been contracted in the lab, and one official went so far as to suggest that the tularemia was incidental, and that the mice had really died from the psychological stress of being confined in Mason jars. Until mice were actually found dying of tularemia, or some other disease, *in the field,* it could not be safely suggested that the end of the mouse plague was in sight. J. D. Vertrees's search had been thorough, but he was only one man working on one day. Perhaps a sustained, large-scale effort . . .

And thus it was that two weeks after the first mouse died in Hamilton, Dr. Jellison came back to the Klamath Basin for what promised to be a long stay. He arrived by air late on the evening of Tuesday, December 3, bringing with him another Rocky Mountain Laboratory physician, Dr. Fritz Bell, and a planeload of equipment that was immediately transferred the short distance from the county airfield south of Klamath Falls to the low cinder block buildings that house the Federal Cooperative Extension Service's Klamath Experiment Station on Henley Road. Here, in a fifteen-by-thirty-foot back room, a full pathological laboratory would be set up. (In 1978, as I was researching this book, Paul Hatchett showed me the room. It serves today as a staff lounge and contains nothing more esoteric than a coffee pot. Such was not always the case. "We had fifty thousand dollars worth of equipment and eight or ten men in here, fighting the invasion," Hatchett told me.) Drs. Frank Prinz and Leo Kartmann of U.S. Public Health's Communicable Disease Center in San Francisco were due the next day, and Dr. M. A. Holmes, Oregon's Public Health Veterinarian, would arrive within a week. All five men would stay until the mice were licked or replacements were sent, whichever came first.

On Friday, December 6, the county agent's office and the five public health experts held a press conference in Klamath Falls to explain their work; and four days later, on December 10, most of them traveled the 280 miles to Portland for an all-day meeting with representatives of the State Board of Health, the Predator Control Bureau, Oregon State College, the Agricultural Extension Service, the State Game Commission, and the State Fish and Wildlife Service (the State Department of Agriculture was conspicuous by its absence). The themes of both the press conference and the Portland meeting were the same: uncertainty, lack of knowledge, and a feeling of being overwhelmed by a phenomenon that no one quite understood. "We have never seen a comparable situation to this one," summed up Fritz Bell, "and not knowing what has happened before, we have no way of predicting what will happen." It was generally agreed that poison programs, despite their attractiveness, would probably be of little value. "I don't think," remarked Vertrees, "that man has the tool that will alleviate the situation." Prinz agreed, adding that nature should be allowed to solve the problem through an epidemic of tularemia or some other disease. Would this happen? Prinz shrugged. "It always has," was the best answer he could come up with. Of course, "always" might not be adequate. "I've never seen anything like this infestation in the Klamath Basin," Prinz admitted.

Many speakers pointed out that since overcrowding was one of the triggering mechanisms for epidemics, a poisoning program might actually be counterproductive: By thinning the ranks of the mice, it might delay the onset of disease and thus prolong the outbreak. However, no one seriously expected a nonpoisoning recommendation to be followed. A rancher being invaded by an army of several hundred thousand hungry mice could hardly be expected to stand by and watch while they helped themselves to his crops. The experts admitted that they were stumped by this dilemma. And that was all that was needed for the politicians to take over.

Shortly after the Portland meeting, Oregon Governor Robert D. Holmes called the press to announce that he was seeking federal assistance for the farmers of the Klamath Basin and other infested parts of eastern Oregon. "California and Nevada are also affected," the governor pointed out, "and it could spread all over the West." Senators Wayne Morse and Richard Neuberger joined the governor in his request: Neuberger specified that he wanted help from the U.S. Department of Agriculture, the Public Health Service, and the Fish and Wildlife Service, a bit of grandstanding that won him very few points. Public Health, after all, was already involved, and Fish and Wildlife had other fish to fry. They announced somewhat grumpily that they had no funds to give Oregon for mouse control, and that, anyway, "wholesale poisoning" would be dangerous to wildlife, and they wouldn't give the money if they had it. The State Department of Agriculture continued to hold itself aloof. In the midst of one of the worst crises ever to hit farmers in Oregon, the departmental director, Robert J. Seward, announced that he was canceling the December meeting of the Board of Agriculture— "due," he said, "to insufficient business."

The *Oregonian,* evidently fed up with the proportion of talk to action that was coming out of all this, published an editorial cartoon showing a mouse, labeled "Eastern Oregon Field Mouse Infestation," leaping nimbly away from a confused-looking cat labeled "control efforts." They called it "Cat and Mouse Game."

Governor Holmes, Director Seward, and Senators Morse and Neuberger were all Democrats; Secretary of State Mark O. Hatfield was a Republican who just happened to be running for Holmes's seat. Therefore it came as no surprise whatsoever when, the day after the Holmes-Morse-Neuberger aid request was made public, Hatfield issued a blistering, headline-grabbing denunciation of it. In seeking federal assistance, Hatfield charged, the Democrats were guilty of "lack of leadership." Oregon had always prided itself on being able to solve its own problems in the past, and there was no reason that the state couldn't solve this one by itself, too. For Seward, a Holmes appointee who had been in office for less than a year, Hatfield had particularly harsh words. "I am shocked," he cried, "to learn that the Chairman of the State Board of Agriculture cancels a December meeting and blandly states his board won't meet again till February because of insufficient business to transact at a time the ranchers and farmers of Lake, Klamath, and Deschutes areas and Sauvies Island are desperate for relief from a plague. . . . He justified his action by saying each meeting costs two hundred fifty dollars, but that is about the tragic cost per minute to those who have infested lands." How could the situation have been better handled? Hatfield outlined a tough, four-point program of his own: It called for (1) a joint emergency session of the State Board of Agriculture and the State Board of Health; (2) formulation of a plan of attack by a team of veterinarians and microbiologists; (3) funding from the state's Emergency Fund; and (4) a series of regional meetings to exchange information with other infested states.

Holmes and Seward reacted strongly to what they saw as interventionism on the part of their upstart secretary of state. Hatfield, Seward muttered darkly to reporters, had been "improperly advised." His department had "no legal authority or responsibility in controlling the mice in southern Oregon." They weren't the state's Department of Agriculture mice—they were federal mice, and Department of Health mice. He refused to discuss the matter further, a refusal which extended into print, apparently as official departmental policy. In the very next issue of the Department of Agriculture's quarterly report, the *Agriculture Bulletin,* there appeared a list of the ten most important plant pests in Oregon for 1957. The mice, incredibly, were not among them.

Holmes's response to Hatfield's sniping was little better. Stung by the secretary of state's statement that Oregon ought to be able to take care of its own affairs, the governor announced—on Christmas Eve—that he was appointing J. Ralph Beck, associate director of the Oregon State College Federal Extension Service, to the newly formed position of "Mice Control Coordinator." Beck was to "prepare a coordinated program with government agencies and the ranchers" to fight what Senator Morse had called "the unprecedented mouse population." This sounded good. It was only when the reader continued, and discovered that a large part of the job was to consist of trying to find funds to do the job, that suspicion began to arise. In truth, Beck's position was nonexistent. It was never mentioned again. A few months later, when an actual working coordinator turned out to be needed, Beck's own agency—the Federal Cooperative Extension Service—hired a young Oregon State College graduate from Nevada named Ed Hansen. The "Mice Control Coordinator" had faded back into the political woodwork whence it came.

Since Hatfield had no authority to enact any of his plan, and since he was running so hard for Holmes's seat the next fall, it is tempting to dismiss the whole thing as partisan politics. Things are rarely that simple, however. It must never be forgotten that partisan politics leads to elections, and that elections lead to terms in office. Whatever his motivations, Hatfield continually gave the impression of sincere, responsible concern. Holmes did not. When Hatfield came to Klamath County to tour the infested area, and Holmes stayed in Salem, that impression was confirmed. It is probably not too much to say that the Klamath mice were at least partially responsible for Hatfield's defeat of Holmes the next November and therefore, indirectly, for Hatfield's later brilliant career as governor and United States senator. God, as the old hymn states, moves in mysterious ways.

8

IN EARLY JANUARY of 1958 *Newsweek* picked up the story, and the resulting groundswell of publicity continued the job that the political brouhaha between Holmes and Hatfield had so effectively begun: It obscured, at least as far as the newspapers were concerned, the most revealing fact that had yet come to light about the irruption. It was this: Those mice had far fewer diseases than they seemed to have the right to. They also had far fewer ectoparasites —fleas, lice, and mites—than should have been expected. In fact, those first fifty mice that had been autopsied by William Jellison and his colleagues in their Montana laboratory back in November had been hosting no ectoparasites at all! Were they actually that much healthier than normal, and should this be considered significant? I asked Walt Jendrzejewski that question, that March morning with him in Klamath Falls twenty years later, and he just snorted. "Healthy?" he said. "Oh, my gosh, yes, BIG fellows! But, yeah, there were far fewer fleas than normal, and we were looking at that pretty hard. . . ."

Parasites and diseases are interrelated because the parasites often act as carriers for the disease, spreading it from one animal to another. The most important of these disease carriers, or vector organisms, are the blood-suckers— fleas, lice, mosquitoes, mites, biting flies, chiggers, and the whole list of other such itch causers. The effects of these creatures and the diseases they carry can often be much larger than people expect. In Texas not long ago, for example, a massive program was launched to fight the screwworm fly, a serious pest of cattle, by releasing sterile male flies to mate with the females and produce infertile eggs. The program succeeded in cutting the screwworm fly population far down, but it also caused a population explosion in the local deer herd. It turned out that the screwworm fly, as well as being a serious pest of cattle, had also been a serious pest of deer, and that removal of it had caused the animals' longevity to shoot up. In the end, the deer caused about as much damage to the farmers' crops as the fly had been causing to their cattle— which meant that one's judgment of the value of the screwworm fly eradication program would have to depend upon whether one was a cattle rancher or a truck farmer.

Not all attempts at vector control have led to such unfortunate results, of course. Many have been extremely successful. Malaria has been conquered in most parts of the world by elimination of the mosquitoes that carry it, with little or no effect on other animal life except in those places where the principal tool of eradication has been such broad-spectrum pesticides as DDT. Similarly, bubonic plague, which is carried principally by lice, which spend most of their lives on rats, has been controlled in Europe and America largely by limiting the mixing of rats and humans; in the Orient, where such mixing is more common, there have been recurrent epidemics of bubonic plague every few decades for centuries.

On the other hand, there have also been some striking examples of success in controlling crop pests by deliberately *introducing* diseases and their vector organisms. The best known of these successes is probably the case of the Great Australian Rabbit Invasion, and it deserves a recounting here in some detail.

Australia has almost no "normal" mammals. Its mammalian life is dominated by marsupials—animals that bear their young while those young are essentially still fetuses, incubating them the rest of the way into young animalhood in a pouch on their bellies. The best known of these is, of course, the kangaroo, but there are more than one hundred other marsupial species in Australia, occupying ecological niches used by the "normal," or placental, mammals elsewhere in the world. There are marsupial mice, marsupial cats, marsupial bears, a marsupial wolf, marsupial squirrels, and nearly every other type of animal that one can think of. Because of the difficulties inherent in the marsupial birth process, all are slow breeders, but this does not matter; the predator population is extremely low and consists mostly of marsupials itself. The single exception to this is the dingo, or wild dog, the only non-human placental mammal in Australia when Europeans arrived toward the end of the eighteenth century. The dingoes appear to have come to Australia as pets of the early aborigines and to have gone wild shortly after their arrival. They hunt in packs, like wolves, running along the ground; since most marsupials can climb trees, and those that cannot are, like the kangaroo, extremely fast runners, the dingoes were not much of a threat, and their numbers remained small.

Enter the European rabbit.

Some time toward the middle of the nineteenth century, a prosperous Englishman named Thomas Austin came to

Australia and established an estate called Barwon Park near Geelong, at the southern tip of the continent. Austin had enjoyed hunting rabbits at home in England, and he was disappointed to find none of the little creatures here. He sent home for some. They arrived in late December 1859, two dozen strong, and were promptly released at Barwon Park. Austin oiled his gun collection and sat back to await the results. They weren't long in coming.

Rabbits, of course, are subject to a whole host of predators in Europe, and in response have developed an extremely high birth rate. Here there were few predators, and this high birth rate simply produced more rabbits. By 1865—barely six years after his precious cargo had arrived—Austin had shot over twenty thousand of the things and barely made a dent in them. There were several million more running about the Victoria countryside. By 1900 they had eaten their way through a third of the continent and were still spreading, and the government had begun to feel desperate. They tried importing foxes: The foxes ignored the rabbits in favor of the slower-moving marsupials, and at least one species—the tungoo—was driven perilously close to extinction. They tried throwing a seven thousand-mile-long "rabbit fence" along the invasion front; the rabbits either leaped it or tunneled under it, and continued to chomp their way across the grasslands. This sort of comedy of errors continued for decades, with nothing that humans could bring to bear against the rabbits' making much of an inroad into their population, until 1950. That was the year a disease called myxomatosis was introduced from South America.

Myxomatosis had coevolved with the South American rabbit, and the South American rabbit had a certain amount of resistance to it. The European rabbit did not. Myxomatosis spread like wildfire through the rabbit population of Australia. Over 99 percent of the animals died. The rabbit problem was over.

Unfortunately, that's not quite the end of the story.

A few years later, the disease was introduced into Europe by a Frenchman named Paul Armand-Delille to control rabbits in his vegetable garden. Armand-Delille was a physician, and he should have known better. The disease spread rapidly throughout Europe, where the rabbit had been in balance with its predators. Lynxes suffered a major population dieback, and those that survived took to attacking deer. Foxes caught more mice, which meant owls and weasels caught fewer mice, and the owl and weasel populations declined. But there were some unexpected benefits, too. Without the rabbits around, reforestation efforts in France, long unsuccessful, began to take hold; so did dune-stabilization programs in Holland. And in England, where rabbits are not native but had been introduced by William the Conqueror, crop yields increased by 50 percent, the growing season lengthened, and several much-loved species of wildflowers, thought to be rare or extinct, suddenly became common. The success or failure of myxomatosis in Europe obviously depends, like that of the screwworm fly program in Texas, on your outlook.

The story of myxomatosis in Australia and Europe demonstrates quite clearly that disease *can* have a strong controlling effect on animal populations. Just how much of an effect it *will* have, however, under natural conditions, is still open to question. Myxomatosis was, after all, in both these cases, a form of biological warfare—an introduced disease that the target animal had no resistance to. Just how much of an effect can one expect in cases where the disease and its target animal have coexisted for millennia?

The answer turns out to be just about what one would expect. The effect of a natural disease on a natural population is not so strong as the effect of myxomatosis on the Australian rabbit hordes, but it is by no means negligible. Studies of field mouse population cycles, for example—cycles like the one that had run amok in the Klamath—have shown quite clearly that in the vast majority of cases, epidemic disease accompanies the population decline. The type of disease involved in the epidemic varies greatly from time to time and from place to place, and can be just about anything virulent enough to kill and catchable enough to be passed easily from animal to animal in a dense population. Charles Elton—the father of a modern population biology, and the first individual to do large-scale work on the problem of vole population cycles—found toxoplasmosis in several successive cycles of field mice during his ground-breaking studies in England back in the 1920s and 1930s, and he and his fellow researchers had just about settled on it as the primary cause of the decline phase of the cycle until it disappeared and the cycles continued. Typhoid is common in the decline phase of the cycles in France; tularemia is often found in American "mouse plagues" and is common enough in migrating lemming populations in the Arctic that the Norse refer to the disease as "lemming fever."

The variety of diseases involved, and the fact that some cycles appear to take place without any epidemic whatsoever, make it quite clear that disease is not a basic cause of the cycles; however, it is equally clear that if the disease shows up at the proper time it can hasten the population crash and thus is almost certainly a controlling factor in the *amplitude* of the cycles, helping to determine how high the "highs" will climb to and how low the "lows" that follow them will descend. However, having this knowledge is one thing, and using it is quite another matter, much trickier than one might suppose. Just how tricky it can be is shown in the strange case of Danysz' Virus, and I think it is worth a little more time away from the events of the Klamath mouse experience to discuss that case now.

Danysz' Virus was not actually a virus at all, but a bacterium of the genus *Salmonella,* closely related to the bugs that cause salmonella poisoning and typhoid fever in humans. It was discovered by a Polish-born French pathologist named Jean Danysz near Charny, a town about sixty miles south of Paris, in 1893. Since Danysz had isolated his "virus" from a dying population of field mice, he concluded that it would be an effective weapon against field mouse infestations; and the French government, anxious for relief from the plagues that seemed to be hitting them every time the field mouse population cycles peaked, or about once every three years, agreed to test this theory. An infested field near Aigre, in the wine-producing region north of Bordeaux, was chosen for the test; surrounded by vineyards, it seemed to represent an island population of mice, in which the influence of the virus could be clearly seen. The field contained an estimated 12,480 mouse holes. They raked it, waited two days, and counted again; 1,304 holes had opened in those two days. Danysz applied his "virus," the field was raked again, and the testers sat back to await results. They waited eight days and counted a grand total of thirty-seven holes. Danysz' Virus was a success. The French government immediately began producing it and distributing it; the world's problems with field mice appeared to be just about over.

Unfortunately, things didn't turn out exactly that way.

Almost at once, it was discovered that Danysz' Virus was actually a variant form of the microbe *Salmonella enteritidis,* the germ responsible for typhoid fever in humans. The variant was too weak to cause much human damage, but there was always the chance that it would mutate back to the pure strain and cause a typhoid epidemic, and no other country besides France would license its use. France itself continued to use the virus sporadically up until the time of the second world war, but with mixed success: Sometimes it would work, and at other times it would have no effect whatsoever. The reasons for the failures are not quite clear, and may have varied from case to case, but it seems likely that most of them were due simply to improper timing. If the microbe were spread about the fields during the decrease phase of the cycle, it would hasten that decrease; but if it were spread about during the *increase* phase, it would merely weed out the less resistant individuals, leaving a healthy population of disease-resistant mice to breed undisturbed right up to the population peak. In such cases, of course— because the breeding stock was healthier and more resistant than normal—the peak phase of the cycle would be likely to be longer and more intense than if Danysz' Virus had not been applied in the first place. Tending to bear this out is the fact, pointed to by Charles Elton, that French vole plagues have tended to be less frequent but more destructive in the twentieth century than at any time before. The cure was worse than the disease, and Danysz' Virus was not just a failure—it was an aid and comfort to the enemy.

Danysz' Virus remains, to this day, the only large-scale attempt known to control field-mouse plagues by means of artifically induced epidemics, and the mixed record of its success does not encourage further experimentation. The sometimes spectacular failures of this artificial disease, however, should not blind us to the importance of the role played by natural disease in suppressing or ending these plagues. We have already seen how naturally induced epidemic disease can play a large role in hastening population crashes during the decline phase of a vole cycle—the toxoplasmosis in Elton's English voles being one outstanding example. Suppose the time were ripe for such an epidemic, and the epidemic, for some reason, did not occur? This could happen if an earlier bout with the disease, such as a mistimed Danysz' Virus, left the voles more resistant than usual; or it could happen if the vectors that should have carried the disease—the fleas, lice, chiggers, mosquitoes, and so forth—were absent. To have an epidemic, you must not only come up with a virulent disease—you must come up with a rapidly spreading one. And if the bugs that carry the bugs aren't there in good healthy numbers, this spreading won't take place. In such a case the vole population would have to continue spiraling upward past the point where disease should have cut it off, until or unless some other factor should intervene.

And this is why it is relevant to ask a few questions about the flea population of the Klamath Basin during the winter of 1957–58. Were flea numbers normal, or were they not?

The answer to this question turns out to be extremely interesting. Flea and other ectoparasite populations on the field mice, it turns out, were not only lower than normal, but were, in many places, nonexistent! Those first fifty mice that Jellison had autopsied in Montana were no anomaly: the failure to find fleas and lice continued through the winter. "During January . . . no fleas were found on these hosts," reads the report of the pathologists from the U.S. Public Health Communicable Diseases Center in San Francisco who were studying the problem.

"*Microtus* nests collected during that period were also devoid of fleas. . . ." Mites, lice, and other vectors were also next to nonexistent; in fact, says the report, "practically no ectoparasites were found," a situation which it admits is "a problem of interest. . . ." Part of the reason for this strange state of affairs was certainly the fact that ectoparasites normally drop in numbers in cold weather anyway. But some could still be found on animals taken from nearby woodlands, so this was obviously not the entire answer. *Some* of the little creatures should have been there. What had happened to them?

The *Oregonian* thought it knew. Back in November, in the same editorial that had suggested the role of predator control in the outbreak—and had thereby drawn the wrath of the livestock interests—the newspaper had also commented on the absence of lice, and suggested a possible reason for it. "That man may be responsible for the mouse 'plague' in Oregon not only through extermination of hawks and coyotes, but also by spray-killing of insects, is indicated," wrote the editors. Noting Jellison's failure to find lice on his autopsied infestation-area mice, the editorial concluded: "Plague is carried by lice which live on rodents and this is a hazard to human health and life. But it would seem to follow that a liceless [sic] mouse is himself a healthier creature, likely to live longer to beget more and larger and healthier litters."

Was the *Oregonian* correct? To find out, we must examine two factors: The susceptibility of ectoparasites to broad-spectrum pesticides, and the spraying levels in the Klamath Basin. If both were high, we have a case. If one or the other is low, we will have to look elsewhere.

The answer to the first of these two questions turns out to be remarkably easy to obtain: One can find it in nearly any textbook on entomology that one would care to pick up. Ectoparasites in general are, indeed, extremely sensitive to the broad-spectrum pesticides—especially the chlorinated hydrocarbons—and these pesticides are often used to control them. One of the most common treatments for scabies, for example, is to dust the infested part of the body—usually the genitalia—with DDT, killing the scabies mites. DDT was also used for many years in American immigration centers to dust down newcomers from Europe and Asia and keep from importing lice, which might be infected with disease, into this country. And fleas are at least as susceptible to the stuff as are mites and lice. At about the same time that the Klamath Basin was experiencing its troubles with mice, the Russian researcher M. R. Netsengevitch published the results of a series of tests in which he exposed fleas to measured amounts of DDT under laboratory conditions. He found that, at the concentrations sold in those days as flea powder for dogs and cats—two grams of actual compound per square meter of surface area—five seconds' exposure was enough to kill half the fleas in a test batch, and that much smaller amounts, as little as 0.1 gram per square meter (the tiniest amount tested), would kill 100 percent of the fleas if they were left in contact with it for just one hour. Clearly, intensive spraying *could* cause these parasite populations to decrease dramatically.

Answering that second question, however, is not quite so easy.

We know, of course, that spraying levels were generally high throughout the nation in those years immediately following World War II. All across America, insects and weeds were being "conquered" to the point where it looked as though for the first time in the long history of agriculture it might be possible to routinely raise perfect crops—bug-free, uncontaminated by nonfood plants, and hugely productive. The tools to win this battle had mostly been discovered shortly before the war, but armies all over the world had seized them for military use, and their general availability to farmers had been forced to wait until the mid-1940s, after the war was over. These were the synthetic pesticides and herbicides: Besides DDT, they included 2,4-D; Malathion; parathion; endrin; demeton: heptachlor; toxaphene; and a host of others. Eventually it would be discovered that the use of these chemicals caused almost as many problems as it solved, and application of them would be cut far back; but that was eventually. Now, in the early and middle 1950s, they were in their heyday: Their use was nearly universal, and new ones were being discovered each year.

It does not follow automatically from this, however, that pesticide spraying was especially heavy in the Klamath Basin. It is at least theoretically possible that Klamath farmers resisted the national torrent pushing everyone else toward chemical farming, and that spraying here was at lower-than-normal levels. There was some pesticide use: We saw that back in chapter 5. But how much was "some," and how did it compare with amounts in use elsewhere? These questions are next to impossible to answer. Records bearing on the matter are apparently nonexistent. The County Agent's office in Klamath Falls has none; neither, as far as I was able to determine, does the central office of the Oregon State University Federal Cooperative Extension Service, the records department of the State Department of Agriculture, the U.S. Department of Agriculture, or anyone else. Concrete proof of intensive spraying—in the form of lists of what was actually sprayed, how much of it was used, and when and where it was applied in the Klamath Basin—simply does not seem to be obtainable.

92

Fortunately, there is another way to approach the problem.

All through the early 1950s, the Klamath County Agent's office published a weekly column in the *Herald and News* called "County Agents Report," and that column commonly gave spraying recommendations. By reading through a couple of years of these columns, it is possible to construct a list of sprays being suggested for use, including names, target organisms, and amounts per acre. Here is such a list, prepared from the weekly reports for the years 1957 and 1958; although total amounts and acreages are still absent, the list at least strongly suggests that high levels of spraying were, indeed, being used in the Klamath Basin for those years. The recommendations:

JANUARY: Apply CMU and chlorate-bordate mix as soil sterilants. CMU, also known as Monuron, is a chemical cousin of 2,4–D and 2,4–5T, the chemicals used by the U.S. Forest Service to control brush and by the U.S. Army—as "Agent Orange"—to defoliate Viet Nam; chlorate-bordate is a mixture of sodium chlorate (for toxicity) and borate powder (to counteract the extreme flammability of the chlorate). Put on in heavy concentrations during the rainy season (CMU 60-80 pounds per acre, chlorate-bordate 500-600 pounds per acre), both substances were designed to be carried into the soil by rain and snowmelt to kill all weeds, breaking down into harmless residues by planting time.

FEBRUARY: Spray all desirable plants with a lime-sulphur mix as a prophylactic against possible aphid infestation. Lime-sulphur was a holdover from the pre-synthetic pesticide days, and can be classed as relatively harmless. February got by easy.

MARCH: Treat clover fields with DDT or heptachlor

(DDT: one pound per acre; heptachlor: one-half pound per acre) to kill clover root weevils. Heptachlor is one of the chlorinated hydrocarbons, a close relative of DDT but with much greater toxicity.

APRIL: Continue the clover-field treatments. Toward the middle of the month, begin spraying dandelions with 2,4–D.

MAY: Continue 2,4–D as a dandelion treatment; later, expand the spraying to include brush patches you want to get rid of. For rabbit brush, three pounds per acre of the herbicide should be applied just as the plants show three inches of new growth; for sagebrush, there is a complex timing schedule involving a separate plant, Sandberg's bluegrass, as an indicator of when to spray. This month, too, you should begin the first of two specific treatments for potato aphids as a follow-up to the general prophylactic that you gave in February. After the new crop has sprouted but before it is six inches high, count the number of aphids per patch. If the number is more than one bug per hundred leaves, treatment will be called for. Use Systox (also known as demeton) at a concentration of one-half to three-fourths of a pound per acre; this is a systemic poison which will render the potato plant itself poisonous to the aphids.

JUNE: Continue the potato-aphid treatment. Also, toward the end of the month, spread a two-pounds-per-acre dose of toxaphene—another DDT relative—to combat lygus bug and seed weevil in your alfalfa and clover.

JULY: Toward the middle of the month, give the second of those two specific treatments to your potato aphids. Systox won't work well—and is dangerous—when the plants have grown this far, so endrin is recommended, at a dose of 0.8 pound per acre; this will also get rid of any army worms or cut worms that happen to be in your potato patch. Endrin is the most toxic of the chlorinated hydrocarbons—the same family that contains DDT, heptachlor, toxaphene, and several others. It is poisonous enough that even in the 1950s, before the great pesticide scares, it was usually recommended for use only in the spring—unless, of course, you had to use it against potato aphids in July.

AUGUST: Late in the month, and on into September, spray your trees and bushes—fruit trees, nut trees, currants, grapes, and most ornamentals—to kill the young of the lecanium scale before they become attached to the plant and develop their hard shells. Use one of the organophosphates, preferably Malathion. Organophosphates are considerably more toxic than chlorinated hydrocarbons, but have the advantage of breaking down rather rapidly in the environment, while chlorinated hydrocarbons may persist for decades. Besides Malathion, other organophosphates include parathion and diazanon.

SEPTEMBER: Continue the lecanium-scale treatment, and spray your morning glory with 2,4–D.

OCTOBER AND NOVEMBER: With your crops all in, and with most insects either dead, dormant, or in egg form for the winter, you can afford to take the time to do a good, thorough job of soil fumigation. Disc ethylene dibromide and/or dichloropropene into your bare fields. These release volatile chemicals as they are warmed by the sun during the coming spring, and the volatiles will rise slowly through the soil and kill all your nematodes.

DECEMBER: Peace on Earth, good will toward men.

In studying this list, we must bear in mind that it is a series of recommendations only—not regulations. Farmers could, and often did, exceed the recommended doses, and use these and other chemicals for other weeding or pest-removal jobs. All the synthetic pesticides were widely regarded as interchangeable: If you didn't have DDT, you could use heptachlor, or dieldrin, or endrin, or parathion, or diazanon, or just about anything else. They were also widely regarded as being more pest-specific than they actually were. Insecticides were supposed to kill only insects (although some county agents recommended a few, such as toxaphene and endrin, as mouse-control tools). Herbicides, similarly, were supposed to kill only plants, and unwanted plants at that, though it could be demonstrated that CMU—to pick one example—would kill small fish at concentrations as low as twenty parts per million (yes, that's the stuff they were using as soil sterilants, at sixty to eighty pounds per acre, in fields that would send irrigation runoff into the fish-filled waters of the Tule Lake and Lower Klamath wildlife refuges).

Accounts by Klamath Basin residents quite strongly bear out the evidence of this list, in terms of what it says about the amount of spraying being done. "The spray planes were buzzing around all the time," recalls a friend of mine—nameless here—who attended Klamath Union High School from 1956 through 1960, "and the pilots had very short life spans. It seems as though they were always cracking up." My friend had a summer job for an agricultural supply firm in Merrill, mixing the sprays to the proper dosages; his hands were "constantly yellow" with parathion powder, and I suppose he has the fact that his skin is quite dry to thank for being alive today. Later he worked for a while at the Tule Lake Irrigation District, pumping Aqualin—a World War II tear gas—into the canals to clear them of moss. "I can't tell you how much we used," he says today, "but the stuff came in fifty-five-gallon drums, and it seemed like I was always wrestling those things around." As the chemical was pumped into the water of a canal, small fish would float to the surface, belly up; a cloud of seagulls always seemed to be following his pumper to pounce on the dying fish. "I don't think the chemical affected the birds at all," he says, "but all that means is, I never saw one fall out of the sky."

It is easy for us today to condemn the reckless manner in which ton after ton of these highly toxic compounds was poured into the soil and onto the plant life of American farms, but such criticism would be unwarranted. *Silent Spring* was still years away from publication; the long-term side effects of the chlorinated hydrocarbons were only dimly understood by a small handful of dissident scientists; and in the heady, Brave-New-World atmosphere following World War II it was inevitable that these products—so superbly effective against crop pests— would be seen only as one more marvelous tool bestowed on humanity by the cornucopia of benificent technology. Even the great majority of biologists and chemists, who might have known better, felt this way. The farmers were not biologists or chemists: All they had to guide them was the county agents' recommendations, the texts of ads placed by firms like the Veliscol Chemical Company of Berkeley, manufacturers of heptachlor ("Agricultural Experiment Station tests prove that Heptachlor increases yields from 10% to 100%, so PLAY IT SAFE!"), and the knowledge that DDT's discoverer, Paul Müller of Switzerland, had just been awarded the Nobel Prize. There is little wonder that the chemical rain was so intense. This cavalier attitude should not be looked on as a sign of guilt. It was really a sign of innocence—the innocence of a boundless faith that believed wholeheartedly that the goods of technology were about to turn this planet into the best of all possible worlds.

Already by the mid-1950s there were signs, in the Klamath Basin and elsewhere, that heavy pesticide use might cause massive biological imbalances, thus creating more problems than it solved. Some of the first to discover this had been the citrus farmers of California's Owens Valley. They had depended for more than fifty years on a species of ladybug known as the Vedalia beetle to control cottony-cushion scale, an insect pest that attaches itself to citrus trees, grows a protective scale with an outer covering that looks like a little ball of cotton, and spends the rest of its life feeding off the tree's sap. The beetles ate the young scales before they could attach themselves; they were relatively cheap, and extremely effective. Then DDT came along, and the growers started spraying. Unfortunately, the cottony-cushion scale proved immune to DDT and the Vedalia beetle didn't. What followed, according to entomologist Paul DeBach, was "an unbelievable population explosion of the scale . . . many groves actually looked as if they were covered with snow." Spraying was discontinued, and the farmers—suddenly much wiser—scoured the nearby countries for Vedalia beetles, paying as much as one dollar apiece for them until the population was reestablished and the scale insect was once more under control.

Similar but less spectacular things were happening closer to home around the Klamath Basin. There was an outbreak of woolly aphids on conifers in the Cascades, which no pesticide seemed able to control. One hundred miles to the northwest, in the Willamette Valley, they were experiencing a Hessian fly outbreak characterized as "the worst in twenty years"—despite spraying efforts—and on Mount Shasta, within view of most of the inhabitants of the basin, a grasshopper outbreak had reached levels of fifty insects per square yard for two summers running, despite massive treatment with aldrin, chlordane, dieldrin, and toxaphene. Four years after the great mouse outbreak, in the summer of 1961, damning evidence of pesticide overuse would surface in the Tule Lake and Lower Klamath wildlife refuges as large numbers of birds would be found dead along the shorelines with what proved to be huge amounts of pesticide residues in their bodies—an incident that Rachel Carson was to make good use of in *Silent Spring*. The water supply for the two refuges, of course, was runoff from the irrigated fields of the Klamath Basin—the very same fields that had experienced the explosion of flealess and louseless mice. The conclusion that the lack of fleas and lice, and therefore the lack of disease, was due to spraying in these fields can only be circumstantial, but it seems compellingly strong.

The silence that greeted the *Oregonian*'s publication of its pesticide theory suggests that the scientists studying the mouse outbreak probably agreed with it. However, this agreement did very little good. Perhaps a little more judicious use of agricultural chemicals might have prevented the outbreak, but just now that was quite irrelevant. The outbreak had already occurred. The question was not really where they all came from: That could be dealt with later. The question was much more immediate than that. They already had the mice. Just what the hell were they supposed to do now?

9

JUST WHAT WERE they supposed to do? No one seemed to know, although there was no end of helpful suggestions. L. A. Peterson of Mist, Oregon, a small town on the Nehalem River in the northwestern corner of the state, apparently agreed with the *Oregonian*'s editorial; he was sure that the mice would disappear if the farmers would "let the predatory animals alone for a while. But, no, they have to kill, kill, poison, trap, any way to get rid of those predatory animals. . . ." However, Mrs. Jean White of Portland disagreed. She suggested that a "fleet of fox terriers" be imported to do the job "without sacrifice of birds or chicks." And the editors of the Albany (Oregon) *Democrat-Herald* disagreed with both approaches. Noting that the Willamette Valley seemed to have a "surplus of stray cats," they suggested that these cats be captured and sent to eastern Oregon to help with "the surplus of crop-devouring field mice." It is not clear, from context, whether the suggestion was made facetiously or if the editors actually thought it might work.

The Carson Factor

No such doubt exists about the solution suggested by a disgruntled hunter in a letter to the editor in the *Herald and News*. Signing himself "A. Noni Mouse," this individual wrote that the only way he could see to get rid of the mice was to "turn the problem over to the Fish and Wildlife Service and the State Game Commission. They will immediately declare the whole mouse area a closed reserve. It will be fenced, guarded and patroled. During that period when mice are most plentiful, no hunting of mice will be permitted. Signs to that effect will so advise all hawks, coyotes, house cats, sea and us other gulls. . . . In no time mice will become as rare as the whooping crane."

Among those directly involved in working with the mice, opinion continued to be sharply divided, with farmers and farm officials continuing to call for massive poisoning programs while most scientists continued to insist that such programs would do little good and might actually end up by prolonging the plague. Caught between were the government agencies, which were being asked for action by the farmers but were also being advised by the scientists. The agencies reacted to this dilemma in typical agency fashion, that is, by following that ancient rule of bureaucracy that says: When in doubt, study. By mid-December 1957, six separate studies had been initiated, by at least four agencies. The California Department of Agriculture had done one; the Oregon State College Federal Cooperative Extension Service had done another; and the Oregon State Department of Health was engaged in a third. U.S. Public Health Service had three concurrent ones going—in Hamilton, Klamath Falls, and San Francisco. And so it went. Governor Holmes and his challenger traded political body blows; the mice ate baseball-diamond-sized circular bare patches in the alfalfa fields, nested in the potatoes, and turned the ditch banks into sponges; the farmers complained; and the bureaucrats studied. It is not on record that anyone suggested getting rid of the mice by hiring someone to read all those studies to them, thereby boring them to death, but the possibility must have been considered. So it is probably no coincidence that on the day after Christmas, less than a week after the announcement of still another study—this one by Alva R. Kinney, a control specialist in communicable diseases from U.S. Public Health Service's Communicable Diseases Center in San Francisco, who was being sent in at the personal request of U.S. Assistant Surgeon General Theodore J. Bauer—a group of farmers and county agent's office staff people met to form themselves into a committee to wage what the *Herald and News* called "All-Out War" on the mice.

Out of that meeting there emerged, for the first time, a coherent, well-thought-out, operable plan for dealing with the mouse problem. At its heart was a coordinated, large-scale poisoning effort, to take place simultaneously throughout the Klamath Basin, in all three affected counties. The bait—grain treated with zinc phosphide, a black, crumbly powder described delicately by one U.S. Fish and Wildlife Service biologist as "malodorous" but which seemed to be extremely attractive to the mice— would be distributed by hand in alfalfa fields, clover fields, and pastures, and along the roadways, railroad embankments, and ditch berms where the mice tended to congregate in wet weather. It was estimated that some 250,000 acres of the basin needed this treatment, but because much of that acreage was unbaitable due to various environmental constraints (including flooding and access problems), a critical core of some 66,000 key acres was settled on to be treated with bait at the rate of ten pounds per acre, a total of 660,000 pounds. Timing was especially crucial: If the baiting was done too early, the materials wouldn't be on hand to do it right, but if it was done too late the green spring growth would have begun in the fields and the mice wouldn't be hungry enough to take the bait. February 15, 1958 was chosen as the target date. On that day, weather permitting, the baiting would begin; meanwhile, there was much to do.

The first need was for funds. At 6.5 cents per pound, 660,000 pounds of bait would cost nearly forty-three thousand dollars. A lump-sum expenditure this high was beyond the pocketbooks of the farmers, but the government ought to be able to come up with it—if the government could be persuaded to stop studying long enough to look in its bank account. The lower the government level, the more likely this was to happen. If they started right at home. . . .

On Friday, December 27, 1957, the farmers' committee approached the Klamath County Court—the county's three-member governing body—and formally requested funds to aid the farmers in their mouse-control efforts. The court promised to look into the matter. Judge Charlie Mack thought that the county could probably squeeze out ten thousand dollars, if the law would allow county monies to be spent for this sort of activity. District Attorney Richard C. Beesley was assigned to research the legal question.

The farmers next approached State Senator Harry Boivin. Requests to the state government for assistance had heretofore produced only studies and campaign rhetoric, but those requests had been directed to Salem—and there were no mice in Salem. Boivin lived in Klamath Falls, and it didn't require a study to convince him of the severity of the problem. He quickly bucked a letter up to the Emergency Board—a peculiarly Oregon institution composed of selected members of the state legislature, charged with disbursing state money during times when the legislature is not in session—and asked for a special meeting on the mouse problem. The same day that the contents of this letter were released to the press, District Attorney Beesley announced his finding that the county court could legally give the money to the farmers to fight the mice; the next day, January 9, 1958, a special five-day meeting began in the central Oregon city of Bend, with representatives from all of Oregon's mouse-plagued counties—Klamath, Deschutes, Wasco, Lake, Harney, and Jefferson—in attendance, along with more than one hundred farmers. Unlike earlier meetings, which had been dominated by agency personnel, this Bend meeting's attendance was largely people who were directly affected by the problem. They quickly came up with a resolution to present a ninety thousand dollar-funding request to the Emergency Board. Under the circumstances, it was unlikely to be ignored. Klamath County could begin mixing bait.

The bait mixing was done in what was called the "poison house," a small building on the grounds of the Klamath Experiment Station, which had been used for this purpose before, although never on this scale. "We mixed an awful lot of bait, just tons of it," recalls Jendrzejewski. "In the beginning we used strychnine, but as things started to get worse we used zinc phosphide. That was black-looking stuff—smelled horrible." Zinc phosphide powder, at the rate of two pounds per hundred pounds of rolled oats, was mixed with a small amount of mineral oil to form a sticky suspension and then poured over the oats while they were being agitated in a power-driven mixer. When the oats-and-poison mixture was a uniform black color, mixing was complete. The process was efficient, safe (provided the operators took the precaution of wearing rubber gloves), and fast. Black-looking stuff was soon pouring from the poison house at the rate of ten thousand pounds daily.

The Emergency Board met at the end of January and rapidly appropriated $110,000 to fight the mice—$100,000 for bait and $10,000 for another goldurn study. Now preparations for the mass poisoning moved into their final stage: obtaining the assurance of coordination, without which the program could not hope to succeed. The poisoning program needed to hit the mouse population one single, gigantic blow, rather than a series of little ones, and it was essential that every farmer and rancher in the basin should understand this reasoning and cooperate with it. The farmers' committee, in cooperation with the county agent's office, began a two-pronged attack designed to reach everyone. Agent Charles Henderson commenced a series of articles in the *Herald and News* explaining the mouse plague—insofar as it was understood—and laying out the mechanics of the coming mass poisoning in a way that every reader of the paper who was at all interested in mice should hopefully be able to understand. At the same time, he and J. D. Vertrees began a round of well-publicized meetings, held mostly in Grange halls, designed to acquaint farmers first-hand with the workings of the poisoning plan. These meetings took place in Malin, Merrill, Henley, Poe Valley, Bonanza, and Lorella. Henderson and Vertrees were also the guests of the Klamath Falls Rotary at a late January luncheon, which enabled them to give the local business community a little bit of understanding of what the farmers were facing. The days began winding down toward target. The Tulelake Grange held a last-minute meeting to coordinate efforts in the California section of the basin with the massive plan going forward in Oregon.

Poison continued to pour forth from the poison house. The mice continued to eat things. . . .

Or did they? According to Henderson, they were "wintering well"; Bill Decker of the *Herald and News* reported that "recently caught specimens appear sleek and fat despite the apparent shortage of food." On the other hand, there seemed to be a number of strong indications that the gigantic mouse population was about to collapse of its own weight. As early as mid-December 1957, the Public Health Service had reported that "several isolated colonies" had been found dying of tularemia; by late January 1958 tularemia had been found in the Klamath Basin groundwater supplies—though not in concentrations large enough to constitute a serious human health hazard—and biologist Donald A. Spencer, a rodent-control specialist from Denver assigned to the plague area by the U.S. Fish and Wildlife Service, was reporting that 80 percent of the mice he examined were showing symptoms of the disease. On February 12, as preparations for the poisoning program were reaching their apogee, a conference in Portland brought together representatives from all federal and state agencies that had been studying the mouse problem; the conferees reached agreement almost immediately on a resolution predicting that at least 75 percent of the mice would die from natural causes by late spring and that, while there would quite likely be a buildup in numbers again over the summer, the population would almost certainly not reach plague proportions again for many years. So the poisoning program would be likely to be wasted effort. "Regardless," the conferees stated, "of what control measures may be taken by man, the mouse infestation will continue until nature remedies the situation."

The farmers' committee, not unexpectedly, disagreed

with this prognosis. "New nests and young mice are prevalent, and the green feed has begun to grow," worried the county agent's office. The mice "are going into the breeding season at high levels," fretted the *Herald and News*. Plans for the poisoning went ahead. February 15 arrived in a downpour. The program was postponed, rescheduled, postponed again . . .

And now, as the mass poisoning program teeters on its edge, I think it is time for a short digression into the disciplines of history and ecology, so that we may better appreciate the irony of what happened next.

10

The reproduction of mice is a most astonishing thing
when compared with other animals both for the number
of young produced and the speed of it. . . . In many places
an innumerable multitude appears regularly, with the
result that very little of the corn crop is left. . . .

Aristotle, *Historia Animalium,* 330 B.C.

T HE KLAMATH BASIN is hardly the first place in the
world ever to have suffered a plague of field mice; in
fact, many such plagues have been recorded, in various
locations, right from the beginning of written history. The
Old Testament speaks of "the mice that mar the land,"
and attributes one plague to God's vengeance on the
Philistines for stealing the Ark of the Covenant from the
Israelites. Diodorus Siculus, Strabo, and Pliny all mention
mouse problems. Herodotus included a long account in his
famous *History* of how a mouse plague determined the
outcome of a battle between King Sennacherib of Assyria
and King Sethos of Egypt: On the night before the battle,
Sennacherib's camp was invaded by what one sixteenth-
century translator described as "an huge multitude of
field mice, which gnawed their quivers, bit in sunder their
bowstrings and the braces off their shields, that in the
morning being disfurnished of their armour, they betooke
themselves to flight, not without the losse of many
souldiers." The Greeks had so much trouble with mice that
they developed a theory that the animals could get preg-

nant merely by licking salt, without the need for copulation; they were combatted by fumigation, by turning hogs into the fields to root them out, by protecting birds of prey, and by praying to Apollo, one of whose many duties was to chase mice (a statue of him as Apollo Smintheus— "Apollo the Mouser"—once stood in the temple at Chryssa, in western Crete).

Mouse troubles also plagued medieval and Renaissance Europe. Stories of these troubles abound; for every legend such as that of Bishop Hatto or the Pied Piper, there are dozens of forgotten but much more believable accounts written into local histories. Here is a typical tale, taken from John Stow's *Chronicles* of 1582:

> . . . about Hallowtide last past [1581] in the marshes of Danesey Hundred, in a place called South Minster, in the county of Essex . . . there sodainlie appeared an infinite number of mice, which overwhelming the whole earth in the said marshes, did sheare and gnaw the grass by the rootes, spoyling and tainting the same with their venimous teeth. . . . which vermine by policie of man could not be destroyed, till at last it came to pass that there flocked together such a number of owles, as all the shire was able to yield, whereby the marsh-holders were shortly delivered from the vexation of the said mice. The like of this was also in Kent.

It should be mentioned that the "policie of man" appeared to have gone backward since Greek times. Stow mentions no specific techniques, but it is known from other accounts that—in marked contrast to the Greek practices of fumigation, hog-rooting, and the protection of predatory birds—plagues in the Middle Ages and the Renaissance were combatted (if that is the right word) by sprinkling holy water on the fields, marching through them with the relics of saints (including, in one recorded case near Rouen, France, in the twelfth century, the head of Saint Valentine), and, when all else failed, asking the Church to excommunicate the little devils.

Stow's Danesey Hundred plague was hardly the last to hit Europe: On through the Age of Enlightenment, down through the Romantic Era, and right up to the present, the plagues have continued, growing—if anything—more regular and more virulent, instead of less, with each appearance. Most have been, in comparison with the Klamath Falls experience, relatively minor—with mouse densities measured in the hundreds per acre, rather than in the thousands—but there have been exceptions. One of the worst took place in the valley of the Rhine in 1822. "The ground in the fields was so undermined in places," wrote the German naturalist I. H. Blasius after observing the situation on the lower Rhine, "that you could scarcely set foot on the earth without touching a mousehole, and innumerable paths were deeply trodden between these openings. On fine days it swarmed with voles, which ran about openly and fearlessly. . . ." The Frenchman Charles Gerard described the same plague as it appeared in Alsace, on the upper part of the river, as "a living and hideous scourging of the earth, which appeared perforated all over, like a sieve." Similar destruction hit the Border Hills, between Scotland and England, in 1872 and again in 1892. The British Board of Agriculture did a very fine publication on the latter, including some exceedingly graphic direct reports from affected farmers whose language, laced with Border Hills colloquialisms, seems somehow to accentuate rather than to hide the horror of the affair. "It is impossible to estimate the damage," reads one typical report, "and unless the mice depart soon, and the grass come again in the spring (what was eaten in 1890-91 has not grown again), the most of the sheep will be to take away and keep away till the mice plague gets past. . . . It is impossible for anyone to believe the ground is so sore destroyed unless they see it. They have missed nothing, everything is cropped to the earth. The future is a terrible looking to." Ten years later, in western France, a major infestation with mouse densities estimated at up to eight thousand per acre struck at least twenty-one departments—what we would call counties—out of the nation's total of ninety-four; and fifteen years after *that,* in 1917-18, the mice struck Germany, decimating crops and riddling pastures over several hundred square miles in that country's northern plains region and almost certainly hastening the end of World War I. All of these plagues were similar to the one that struck the Klamath Basin in the 1950s, but this last-named German plague was almost uncannily so. The mice were first noticed in large numbers in early summer, after a mild dry winter; the highest numbers were reached in the fall; the most serious damage was noted in the clover and potato crops; there was a massive poisoning program; and the results of that program were exactly like those that we shall soon see for the Klamath.

Except for the references cited from the Bible and from Herodotus, all mouse plagues we have discussed so far took place in Europe. However, the phenomenon itself is by no means limited to Europe; all Northern Hemisphere continents have suffered infestations. The Asian steppes, those vast, rich plains that form the breadbasket of the Soviet Union, have suffered recurrent plagues, with major outbreaks appearing to take place approximately every ten years; two of these in the late nineteenth century were the worst ever recorded anywhere. North America has also had its share, with occurrences recorded as early as 1699 near the city of St. John, Newfoundland, located on the Avalon Peninsula (sometimes known as the Island of St. John); a second outbreak there, in 1775, was severe enough that many of the residents were reported on the verge of starvation. Other important North American plagues have been reported from Newfoundland (1815); British Columbia (1897–98); Saskatchewan and Alberta (1900–01); Nevada (1906–07); the Valley of Virginia (1918); Manitoba (1922); and Kern County, California (1926–27).

The Kern County outbreak is especially worthy of note, both for its size (it was the largest on record in North America until the Klamath Basin plague came along) and for what turn out to be several unusual aspects. The outbreak was centered on the dry bed of Buena Vista Lake, a former flood-control reservoir, a few miles southwest of Bakersfield in the extreme upper end of the San Joaquin Valley. The lake had been drained only a short time before, and the central portion, where the crops were planted—eleven thousand acres of them, mostly kaffir corn and barley—was not yet fully dry and therefore provided ideal conditions for field mice, which like the soil slightly moist. These ideal conditions were a biological island, surrounded by fifteen thousand acres of alkali flat. There were few predators, both because of predator control and because of the barrier imposed by the alkali flat—which also proved an effective barrier to the spread of disease. And the mouse population (including a healthy component of feral house mice, *Mus musculus,* along with the more common field mice) exploded. By late November, 1926, they had reached an absolutely extraordinary density, estimated by at least one reliable observer at up to eighty thousand individuals per acre (slightly under two per square foot) and begun what turned out to be a series of three major migrations outward across the alkali flats in search of food and space.

Field mice are highly territorial animals, and even under plague conditions they will not normally travel more than a few yards away from their birthplaces, so these migrations—which reached distances of up to twenty miles from the lake—represent a behavioral aberration that is probably unique in the history of mouse-plagues. The migrations were numbered at up to a million animals each, and they moved over the ground in waves, devouring nearly everything—including, in at least one reliably documented instance, a full-grown, living, and apparently healthy sheep that had the misfortune to be penned in their path. The main highway leading north from the lake had a dead mouse for every square yard of surface over a distance of some seventeen miles, and had to be posted for slippery conditions. The Klamath Basin infestation was bigger—in terms of total numbers of mice, infested acreage, and dollar damage—but it is probable that no mouse plague anywhere has ever matched the horror of what the citizens of Kern County had to go through during that winter of 1926–27. Considering the major role that predator control appears to have played in causing this outbreak, it seems rather strange that no lessons were learned and they had to go through the whole thing all over again in Klamath County thirty years later. But then, the lesson does not seem to have been learned in Klamath County, either.

The basic, underlying cause of all this destruction is clearly the vole cycle, or microtine cycle, which we discussed briefly back in chapter 3—that massive, slow rise and rapid fall in the field-mouse population that sends the numbers of mice in an acre of grassland from three or four up to three or four hundred and back down again in what works out to be a remarkably steady three-year rhythm. The reasons for this steady, constant heartbeat of mice over the centuries are not fully understood. It is often considered to be caused by some dynamic interaction between the mice and their predators, but most biologists today tend to discount that notion. "If you go into this thing with the preconceived notion that predators control prey, I can't give you much information," Steve Cross warned me quite early in my discussion with him in his laboratory at Southern Oregon State College. "You see, there's at least four credible theories regarding population fluctuations, and the predator-prey relationship is only one of them. I guess we can start with that. Predators and prey, the theory goes, are hooked into what computer people call a negative-feedback loop. When plenty of prey is available, predators increase their population until they grow so numerous that they overeat the prey and its numbers drop back. This forces the predators to either starve or leave, so their numbers drop back too, and they relax the pressure on the prey, which springs back up again. That theory used to be quite widely held, and still is in some quarters. And it's been proved, I think pretty definitely, that there *is* a cyclic relationship between predators and prey. But it's cause and effect that people worry about. Not whether you can see a relationship, but which is cause and which is effect. And that is by no means as clear.

"Number two is the nutrient-recovery theory, which holds that the nutrient quality of vegetation is cyclical, for whatever reason, and that affects the population. Basically, it's supposed to work something like this: When the plants have a high nutrient quality, the reproductive rate of the rodents feeding on them goes up and the population increases. This ties a lot of the nutrients up in rodent bodies and fecal material, where it can't be used by the plants, and so their quality goes down, causing a higher death rate and a lower reproductive rate until those bodies get recycled into the soil again. And you can't rule that out, because research shows that nutrient quality can definitely influence reproduction. And there *are* cycles in plant nutrient quality. But there haven't been any studies that show much cyclical correlation between the two, and not many people today would tend to hold that theory.

"The third theory," Cross went on, "is what we might call, and I'm looking for a word here, the—well, I guess 'sociopsychological theory' would come as close as any. This one's been demonstrated in laboratory animals, but it does not show well in the wild. Stress by overcrowding is reflected in reproductive abilities; the more crowded populations respond by failing to breed, and their numbers go back down. John J. Christian is the one that has done the most work on that; most of his work was done in the fifties."

I looked up some of Christian's work later. He was a student of the great pioneering population biologist, Charles Elton, in England, and his work, like that of his mentor, is characterized by lucidity, thoroughness, and admirable scientific dispassion. He has proved, I think quite thoroughly, that caged vole populations are affected by overcrowding to the point where they stop breeding, and that artificial population cycles may be created in this way. But it's a long way from caged voles in the lab to free ones in the wild, where there are no constraints on movement and the animals can seek out corners of the field where the population is less dense. And that brings us to Cross's theory number four. What, I asked him, was theory number four?

"Well," said Cross, reaching for a pencil and paper,

"number four is the most current and widely held, and that's the genetic-dispersal theory. And it's a little bit more complicated than the others. It holds that there are two basic groups in the population: those that are aggressive, but with a low reproductive capacity; and those that are nonaggressive, but with a high reproductive capacity. And there's a constant dynamic fluctuation between these two groups. You start with a population of the nonaggressive, highly reproductive animals, and they're gradually replaced by the aggressive ones, who drive the nonaggressives away, so that they disperse outward—that's where the name genetic-*dispersal* comes from—the reproductive rate of the whole population goes down, the population falls, and the nonaggressives have a chance to migrate in and start the cycle over." I watched, fascinated, as a set of diagrams grew on his paper. They appear on page 118.

Note how the dispersal of the N's in figure 2 leads to a much less dense figure 3—and how, as the number of A's drops in figure 4 (because of the low reproductive rate of the A's), there is room for the N's to come back in and create a density much like that of figure 1. It all seems neat and foolproof—except for two nagging little questions. "Why," I asked Cross, as I looked up from his diagrams, "should there be a correlation between aggressive behavior and low reproductive rates? I should think you'd want to ask that question immediately."

Cross nodded, "Well, yeah, and that's a question that hasn't been fully answered," he admitted. "But studies have definitely shown behavioral and genetic differences, both, between the dispersers at the low point of the cycle and the animals at the high point. So there's some real evidence; we just don't know what it means yet. I mean, we can see definite genetic differences, but we don't

```
N   N   N   N
N   N   A   N
N   N   N   N
N   N   N   N
```
Figure 1

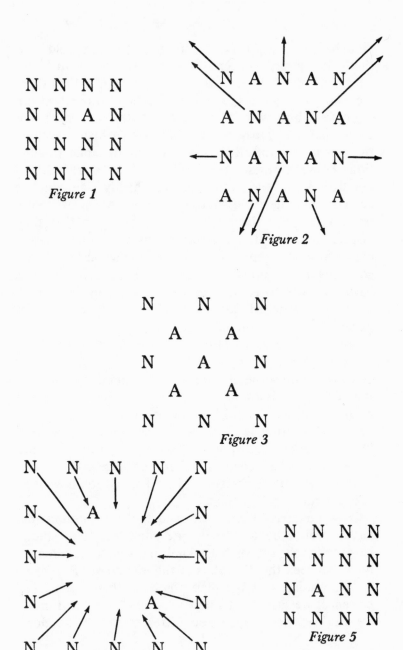

Figure 2

```
N       N       N
    A       A
N       A       N
    A       A
N       N       N
```
Figure 3

Figure 4

```
N   N   N   N
N   N   N   N
N   A   N   N
N   N   N   N
```
Figure 5

know for sure whether we're looking at the right genes or not. What's your other question? You said there were two."

"Well, yes, I did. What is the relationship between this theory—or *any* of the theories, for that matter—and the events in the Klamath Basin in 1956–57?"

Cross grinned at me. "Irruptions," he said, "seem to arrive at a time when the high point of a rodent cycle corresponds with optimum conditions in the environment. But we really have very little information on those. I guess what I'm saying is, you're pretty much free to make up your own theory, and as long as it's reasonable no one can really dispute you." He grinned again. "Come back and tell me when you've got something."

"*You're* a lot of help," I said.

But Cross could tell me no more.

11

WHAT IS THE cause of irruptions? What is it that causes the high end of the normal three- or four-year vole cycle to get so far out of whack that it floods the countryside with small squealing bodies? Why should this cycle—so remarkably steady so much of the time—suddenly take off like a steam engine without its governor, going faster and faster until it blows up in a mushroom cloud of mice and scatters the little creatures so thickly about a place like the Klamath Basin?

Science, Steve Cross tells us candidly, doesn't know. But science does offer a few clues.

In the first place, it seems likely that, whatever the cause of the irruption, it will turn out to represent a simplification of the field mouse's environment. The evidence for this is indirect, but it is quite strong. It is known that the farther north you go in the range of the mouse, the more pronounced the basic cycles become, with higher, much more clearly defined highs and lower lows—each cycle, in other words, more closely approximating an irruption. It is also known that the northern environment is much simpler than the southern, with fewer species and shorter food chains. Most biologists who have studied the microtine cycles agree that the simpler environment is the greatest factor in making the cycles more prominent. "The simplicity of the northern community is seemingly partly responsible for its instability," is the way one currently popular textbook, Terry A. Vaughan's *Mammalogy,* puts it. ". . . In tropical habitats, by contrast, there is an enormous diversity of plants and animals . . . the heterogeneity of the environment, the diversity of carnivores, the intricate patterns of niche displacement and potential competition, and the relatively small percentage of the energy resources available to any one species provide a buffer system against population outbreaks by any species." It seems clear that temporary and localized simplifications of the environment ought to have the same effect on the mouse cycle that the permanently simpler ecosystem of the North has—they ought to make the cyclical highs and lows more pronounced, an effect that we would see as an outbreak.

Secondly, the fact that mouse plagues like the one that hit the Klamath Basin are limited almost exclusively to areas where intensive agriculture is being practiced strongly suggests that the agriculture has something to do with causing them. This was recognized at least as early as 1909 with the publication of S. E. Piper's now-classic study of the 1906-07 Humboldt, Nevada plague for the U.S. Department of Agriculture. "Agricultural development," wrote Piper, ". . . distinctly increases the danger of plagues by furthering the destruction of their [the voles'] natural enemies, by furnishing a great abundance of food, and by increasing the area in which they find favorable homes. The reclamation of arid lands affords most suitable conditions in large areas which were formerly uninhabitable." The same point was made by Charles Elton. "The rarity of such absolute erosion of the vegetation in any natural community of plants and animals," he wrote in *Voles, Mice and Lemmings: Problems in Population Dynamics*—the book that may be said to have created the discipline of population biology—"reminds us that we are witnessing still another situation brought about by the creation of artificial conditions." The connection between agriculture and mouse plagues is not difficult to determine: It is simply a special case of our earlier statement about simplification of the environment causing irruptions. Agricultural activity almost always results in a simpler environment than would have been present under natural conditions. The diversity of native plants is replaced by large reaches in which the vegetation consists of one plant—the crop—only; wild animal numbers typically are cut far back, and the variety of species present is much reduced. This simplification reduces the stability of the natural community in the farming region, and the effect of a change in any one factor—a reduction in the number of predators, an increase in available habitat, a decrease in rainfall, or any of a host of others—is correspondingly magnified.

Finally, it seems quite clear that the immediate cause of any one given mouse plague—the cause that tips the balance and makes mouse population dynamics, in the already simplified system, go haywire—is likely to vary from place to place and from time to time. Common sense will tell you that. If you begin with a number of control factors depressing a population, and then begin reducing or eliminating them one by one, those that remain must bear a greater and greater share of the burden, until there comes a time when each of the control elements remaining is bearing so much that no single one of them can be reduced with-

out overburdening the others. A table with six legs still
may be able to stand after you've removed three of them,
but when you take out that fourth leg—no matter which
of the remaining legs you choose—that table is going to go
over.

What we must do, therefore, is find out what change or
changes occurred in the Klamath Basin, immediately prior
to the mouse outbreak, which might have caused the
population to explode so greatly. We know that we are
looking for a change from previous conditions, because
if everything had been normal in the ecology of the
Klamath Basin, the number of mice would have been
normal as well. We know, further, that the change was
almost surely a simplification, and that it had to be in
some element of the basin's ecology that touched the
mice, either directly or indirectly. It would be useless,
for example, to try to correlate the increase in mouse
density to the increase in the number of cars with wrap-
around windshields on the streets of Klamath Falls.
These two increases were taking place at the same time,
there in the late 1950s, but there is no way on God's
green earth that they could possibly have been related.
Mice have nothing to do with windshields.

What *do* mice have to do with?

The world of a field mouse, like that of any other
living thing, can be divided into the two broad areas of
genetics and *environment*. Genetics represents the inborn
traits, the influences handed down from ancestor to
ancestor through the medium of DNA—deoxyribonucleic
acid, the material that carries the genetic code in all
forms of life from one-celled plants right up through
humans. Environment represents the external traits, the
factors acting on the bundle of DNA-induced characteristics after that bundle has been formed. The two are
separated from each other by the moment of conception.

All characteristics carried in the egg of the mother and the sperm of the father are genetic; once they unite, any further changes in the animal can only be caused by its environment, in the womb or out of it.

Were the changes that caused the Klamath mouse explosion genetic, or environmental? It is not certain, of course—nothing in ecology is ever absolutely certain—but the evidence appears to be overwhelmingly against genetics. We know, from what Steve Cross has told us, that there are differences in the genetic makeup between mice in dense populations and mice in scattered populations—but we also know that these differences are regular and predictable, occurring on the average of once every three years. To have been responsible for the Klamath population explosion, a genetic change would have to meet two criteria: It would have to be abnormal, operating outside the standard genetic flux; and—because of the mathematical impossibility of all those mice springing from a single set of parents within the time frame of the irruption—it would have to take place simultaneously in a large number of separate individuals. The only known mechanisms that can produce simultaneous abnormal genetic changes are radiation and certain somewhat exotic chemicals known as mutagens. Radiation may be immediately dismissed; there is no known source in the Klamath Basin. Mutagens were present in some of the pesticides and herbicides being sprayed on the basin's fields, but these chemicals have been shown to induce *random* genetic changes— and whatever it was that caused the mouse population explosion had to be uniform, not random, in its effect. Besides, genetic changes tend to be permanent, and there was nothing permanent about the population patterns that led to the irruption. Genetics, it seems, may be safely dismissed.

That leaves environment.

125

The Carson Factor

Environment is a big subject, and ecologists normally like to break it down a little further. There are, they say, four basic components of an animal's environment: These are *food* (what the animal eats); *shelter* (where the animal lives, including both the home it sleeps in and the area it roams over while awake); *weather* (rain, sun, snow, wind, temperature and humidity changes, and so forth); and *other animals* (any other creatures, large or small, which interact in any way with the animal being studied). Of these, the first, food, may be safely dismissed as a cause of the Klamath Basin's mouse troubles. It was relatively abundant, compared to the mouse population, at the beginning of the irruption, but it always is at the beginning of any population cycle; its nutrient value may have been abnormally high (a contention that cannot be proved one way or the other), but correlations between plant-nutrient highs and mouse population highs have happened before, without the disastrous results seen in the Klamath in 1957-58. As Cross has pointed out to us, plant nutrient cycles and microtine population cycles don't seem to bear much relationship to each other. It doesn't seem as though examining the food factor would lead us much of anywhere.

The second component, shelter, may also be dismissed, although not quite so easily. Shelter—or, to use its technically more exact synonym, *habitat*—can have extremely large effects on animal populations, and some of these effects are quite well known. Redwing blackbirds, for example, need cattail marshes for breeding grounds, and their numbers go up and down in close synchrony with the number of acres of marshland available each spring. Sturgeon breed only in running water; dam a river, and its sturgeon disappear. Cougars need hunting territories of a minimum of twenty square miles each, and they like

126

the deep forest; cut your forest size down to below one hundred square miles, and your cougar population will drop to under five animals, too small to form a viable breeding unit. These examples could be extended indefinitely. Animals need proper habitat, and their numbers will always tend to increase to the habitat's limits.

For field mice, the proper habitat is moist grasslands—"not flooded," Cross emphasizes, "but moist. If it's too damp, they won't make it. If it's too dry, you'll only get a few borderline individuals." This habitat, it turns out, is exactly what is produced under the system of irrigation practiced in the Klamath Basin in the mid-1950s. "Used to be," says one biologist who is well acquainted with the area, "they'd flood the fields in the winter, pump the water off in the spring, and that was the end of the irrigation. Then the mice would move down off the dikes into the center of the fields and breed all summer pretty much uninterrupted." It was this situation that caused U.S. Fish and Wildlife Service biologist Donald A. Spencer to tell the Klamath Basin farmers during the irruption that they would "have to live with the mice for the rest of your lives, because you raise them on your own places." But Spencer missed the point. There was nothing new about this mouse habitat; it had been there ever since the lakes were drained, and the farmers were already dealing with it every year.

Tule Lake had begun draining in 1912, Lower Klamath in 1917. In 1920, during the course of a normal cyclical "mouse year," the expanding field mouse population had taken advantage of all that nice new habitat the farmers were so obligingly opening up for them. There had been an irruption. It was not as big as the great irruption of 1957–58, but it was pretty impressive nonetheless. "Mice were so thick," wrote Klamath Falls resident Alice Gay a number of years later, "that Mr. Gay pounded out tin cans and literally covered the pantry. All foodstuffs were kept in tin containers. The rodents even ate holes in the paper on the walls by the chimney and dropped down into the dining room. I can still see their little black eyes. . . ."

In the end, of course, they had been forced to resort to poison. Hulled oats laced with strychnine were scattered along the ditch banks during the spring flooding season, when the mice had to leave the inundated fields for the higher ground along the ditches. It became an annual ritual: As soon as the fields were flooded—a part of the irrigation cycle that would take place in any case—the farmers and their hired men would walk the ditch banks scattering poisoned grain, and in this way the mice were controlled. There had been no further habitat changes to speak of since the completion of the Modoc Project in 1941, and the poisoning had stabilized. The "shelter" part of the food/shelter/weather/other animals environmental equation had been adapted to and was holding steady, there in the mid-1950s. It does not appear that it can be blamed for the great irruption.

That brings us to weather.

Of all the factors affecting animal populations, weather probably rates as the most frustrating for population biologists to deal with. There are so many variables involved, and the results of small changes in them can be so striking, that the problem of making sense out of them tends to overwhelm the researcher. Some of these factors are obvious: wind, temperature, rainfall, cloud cover, snow cover. Others are much subtler.

At Tule Lake National Wildlife Refuge, shortly after I became interested in the Klamath Basin's mouse troubles, I paid a call on wildlife biologist Ed O'Neill. A spare, stringy, bespectacled man, an indeterminate number of years past forty, Ed is chief biologist for the refuge complex of the Klamath Basin; he occupies an office in the main building at refuge headquarters, a cluster of low white frame structures tucked up against the eastern base of Sheepy Ridge just above what used to be the high-water line of the old Tule Lake. There is a New England-like taciturnity about Ed at most times, but it broke wide open when I explained the purpose of my visit. "Cause of mouse cycles?" he boomed. "Why, that's easy! You don't have to look any further than sunspots!" I must have looked incredulous, because he chuckled. "You know," he said, "a lot of people believe that today, they're really tied into it. A lot of animals are cyclic in nature, and for lack of something better a lot of folks, even teachers, peg onto sunspots as the reason." I looked up a number of sources later, and discovered that O'Neill was right. Though Charles Elton showed nearly a half-century ago that sunspot cycles and vole cycles were as often opposed— "right out of phase," to use his words—as they were coincident, there are still people today trying to prove the influence of sunspot cycles on small-mammal populations, and as late as 1961 ecologist S. Charles Kendeigh found it necessary to devote more than one thousand words in his college-level textbook, *Animal Ecology,* to refuting the sunspot theory.

But if not sunspots, what? The latitude of possible answers is nothing short of astounding. Sunspots, Kendeigh pointed out, are only the most obvious of the sun's changes, and dismissing them does not rule out the possibility that some other cyclical change in solar radiation may be responsible for the field-mouse population fluctuations. Another possibility seems to be the movement of high- and low-pressure cells in the atmosphere. Working at

about the same time as Charles Elton, a Scottish meteo-
rologist named A. H. R. Goldie discovered that, for Great
Britain at any rate, there was a long-term cycle of predominantly high-pressure weather systems alternating with
predominantly low-pressure ones, and that the period of
this cycle—three to four years—corresponded quite nicely
with the period of the microtine cycle. However, the
mechanics of a system that would connect such seemingly
unrelatable phenomena is by no means clear, and to the
best of my knowledge Goldie's work has never been
followed up.

Another researcher, F. A. Pitelka, made a series of
studies at Point Barrow, Alaska, during the late 1960s and
early 1970s, showing a strong linkage between snow depth
and lemming reproduction: If the snow was eleven inches
deep or deeper, breeding would be high, but if it dropped
to below eight inches breeding would be low. Pitelka
suggested that the functional intertie between snow depth
and breeding rate was probably the insulating quality of
the snow, with a thicker blanket of the stuff keeping the
animals warmer through the cold Arctic winter. Since
lemmings, like voles, are microtine rodents, and since the
cycles of the two are broadly coincident over most of the
Arctic, Pitelka's data is relevant to the vole cycle as well.
Unfortunately it applies only to the Arctic; south of there,
there seems to exist no correlation whatsoever between
snowfall and small-mammal breeding rates.

There are many other possibilities. Elton himself discovered, for example, that he could vary the breeding
rate of his laboratory voles by varying the length of the
period during which they were exposed to daylight each
day: the more hours of daylight, the higher the breeding
rate. He also found that his voles would breed all winter
if he kept their laboratory warm, but if he put them where
they were exposed to winter temperatures they wouldn't

breed unless they were given "winter food"—dried seeds and fresh winter-sprouting grass. Fifty years later, the reason for this odd little quirk of behavior was pinned down by Norman C. Negus and his colleagues at the University of Utah in Salt Lake City. Working with captive populations of *Microtus montanus*—the animal that had caused all the damage in the Klamath Basin—Negus managed to isolate a substance from new grass shoots that acted as a reproductive stimulator. This "vole aphrodisiac" (my term, not Negus's) disappears as the grass ages. What Negus's work means, of course, is that weather conditions that favor the sprouting of grass will also favor the reproduction of voles—provided that Pitelka's snow cover, or Elton's short daylight hours, or Goldie's meteorological cycles, or even O'Neill's sunspots, don't countermand the grass blades.

As you can see, the whole subject is confusing, and often even contradictory, in nature. It is possible, however, to make some small amount of sense out of it. One key is found in a statement that, in one form or another, appears in virtually all accounts of the great mouse plagues of history. This statement is that *the preceding winter was mild.* It may have been mild and wet, or mild and dry; it may have been followed by a winter of heavy snow, or a winter of almost no snow at all, while the plague was actually in progress; it may have set in early, or it may have set in late; but it was almost invariably describable, in some sense or another, as mild.

What can possibly be held in common between a wet mild winter and a dry mild winter, an early mild winter and a late mild winter, a mild winter followed by a heavy one, and two mild winters in a row? There is only one reasonable answer. During that winter preceding the plague, temperatures must have been higher than normal.

In this light, it is interesting to compare temperatures

131

in the Klamath Basin during the winter of 1956–57 to the averages for that time of year in that part of the world. For the critical period of November through February, they are: November 1956, mean temperature 38.8 degrees (average 38.4); December 1956, mean temperature 32.9 (average 31.1); January 1957, mean temperature 25.1 (average 28.7); and February 1957, mean temperature 35.3 (average 33.3). In three out of the four months, the temperature was higher than normal—half a degree in November, two degrees in December and February. Only in January was the month-long average below normal, and even that figure is misleading. Most of the month was *higher* than normal, but a cold snap accompanying the Inauguration Day snowstorm skewed the month's figures out of shape.

Don't be misled by the small size of those temperature differentials: They are more important than they look. After studying the effects of weather on his English voles, Charles Elton commented on the striking importance of tiny differences. "There was a very delicate equilibrium here," he wrote, "and . . . quite a small variation either in the length or intensity of breeding, or in the survival power of old or young, would cause surprisingly large changes in the equilibrium." Temperature changes can effect both breeding rate and survival power. It may help you understand how sensitive these are if you realize at what an extraordinarily rapid pace these animals live. They have a gestation period of just three weeks, are sexually mature at a month and a half, and have a life expectancy, in the wild, of barely eight and one-half months.

What is the mechanism connecting warmer-than-normal weather preceding a field mouse plague to higher-than-normal mouse populations during the plague itself? It seems probable to me that no single mechanism will be found, but rather a group of them. If the weather is wet along with the higher-than-normal temperatures, for example, the sprouting of new grass would be encouraged, leading to higher consumption by the voles of Negus's reproductive stimulator and greater-than-normal reproduction. If the weather is dry, there will be fewer clouds, which means that sunlight, on the average, will come earlier and last longer, and Elton's daylight factor will increase the voles' reproductive rate. If the winter is early and mild, the rains will come in the fall, the grass with the stimulator in it will have a longer growing season, and the voles will have a longer period under the influence of it. If the winter is late and mild, temperatures will be warm longer into the autumn and the preceding year's breeding season will be extended, allowing the voles to go into the winter with higher than normal numbers, which are then increased at a faster-than-normal pace by the mechanisms outlined for warm/dry winters and warm/wet winters. It almost seems a case of heads-I-win-tails-you-lose. As long as the temperatures are above normal, some combination of factors is going to increase field mouse reproduction.

But this leaves us with two more questions. If the breeding rate of voles invariably goes up in a warm winter, why is it the *next* winter that the plague occurs? And why don't *all* warm winters lead to plagues of field mice?

The answer to the first of these two questions is relatively easy. To see what it is, you only have to recall Paul Hatchett's comment that "the stuff got rampant, the wildlife really hit" at first haying time, near the end of June. In all the great mouse irruptions of history, the large numbers of mice are first noticed in the late spring or early summer. These are the previous winter's mice, given a good healthy running jump on the normal breeding season and as a consequence breeding to higher numbers than normal all year.

The second answer is a little trickier, but it is not really difficult either. All you have to do to understand it is to grasp the difference between *necessary* and *sufficient.* To get a ripe tomato out of your garden, for instance, it is necessary to plant a tomato seed. But it is not sufficient *only* to plant a tomato seed: You must also see that the seed is planted in the right type of soil at the right time of year; that the plant gets enough water and sunlight while it is growing; and that the ripening tomato is protected from bugs, birds, skunks, frost, and so forth, until it is ready. Planting the seed is necessary, but it is not sufficient. Similarly, a mild winter is necessary for a field mouse irruption to develop, but it is not sufficient. If the mild winter comes during the decrease phase of the basic underlying vole cycle, for instance, it can slow that decrease down but it can't turn it around; the decrease will still be a decrease, and no irruption will occur. And even if it happens during the increase phase, other factors—predation, disease, poisoning programs, reduction of habitat, and perhaps some others that we haven't considered—may intervene to keep the population from getting truly out of hand. A mouse plague is a relatively unusual phenomenon; a mouse plague the size of the one that hit Klamath County is even more unusual, and has in fact happened only three or four times in history. It must take an unusual combination of circumstances, indeed, to bring such an uncommon event to fruition.

And that brings us, finally, to the topic of *other animals.*

Like the environment in general, "other animals" is a large subject that is normally broken down, for purposes of study, into several smaller units. As in the case of the environment, too, this breakdown is traditionally done in four parts: predators, parasites, pathogens, and competitors. Predators are those animals that eat the creature being studied; parasites are those that get their food from the study animal without actually killing it; pathogens are disease-causing organisms; and competitors are all creatures —those of the same species as the study animal, as well as those of other species—that utilize the same food or habitat and will therefore be competing with the study animal for some limited resource they both desire. Of these four, the last—competitors—can be safely ignored as we search for causes of the great mouse irruption of Klamath County. Competition tends to have an inverse effect on population; the greater the degree of competition, the lower the population of any one of the competing species. This means that we would have to prove an *absence* of competition, not an increasing presence of it, to account for the increasing mouse population. And we can prove no such thing. In fact, quite the contrary seems to be the case: As we showed in chapter 6, the populations of deer, elk, porcupines, and pocket gophers—all of which may be considered at least partial competitors of the field mice for food—were increasing in the Klamath Basin right along with the mice, and there was a plague of grasshoppers (another grass eater) as close as Mount Shasta. The mice were also competing with each other, and the higher their numbers went the fiercer that competition got. All this should have had a damping effect on the mouse explosion, rather than adding more fuel to it. We will have to look elsewhere for the source of that fuel.

Parasites and pathogens are another matter. We covered this subject fairly thoroughly in chapter 8, but it won't hurt to go over the main points again briefly here. Parasites and pathogens are related, because the parasites—

especially the ectoparasites, those that live on the outside of the host animal and can thus pass freely from host to host—serve as carriers for the pathogens and are instrumental in spreading disease through a population. Chief among these carriers, or vectors, are bloodsucking insects such as fleas and lice. We know that fleas and lice were almost totally absent from the mice during the irruption, but were present on animals in nearby woodlands; since the fields the mice live in were heavily sprayed with pesticides and the woods were not, and since fleas and lice are known to be highly susceptible to pesticides, it seems safe to assume that the spraying caused most if not all of this strange lack of ectoparasites. It also seems safe to assume that this lack of ectoparasites was responsible for the failure of epidemic disease to occur, by eliminating the vectors through which the disease could be carried from mouse to mouse. What we don't know—and have no real way of knowing—is how serious this lack of epidemic disease really was. There is a gut feeling that if an epidemic had started soon enough the population would not have reached the terrible densities that it did, but there is no way to prove it. Population crashes have occurred, Charles Elton tells us, without any evidence of epidemic disease—and if that is so, you cannot blame the lack of a population crash on the lack of disease. The best you can say is that it was probably a contributing factor. There had to be something else wrong with the environment, too.

And that brings us back, once again, to the subject of predators and predator control, touched on previously in chapter 6. Was the Great Coyote War responsible, in any way, for the Great Mouse War?

To some—including Dayton Hyde—the answer is clearly "yes." "I'm satisfied in my own mind," Hyde told me, "that the initial surge of mice was due to predator mismanagement, though once they got to snowballing there wasn't any way that any predator was going to control them." This view finds considerable support in the writings of authorities such as Hope Ryden (*God's Dog*) and Rachel Carson (*Silent Spring*). "Without coyotes to hold them in check," writes Ryden, "rodent populations become a problem." Carson says it even more clearly: "The stockman's zeal for eliminating the coyote has resulted in plagues of field mice, which the coyote formerly controlled."

However, there are those who disagree. "A lot of people say it was killing off all the coyotes, and shooting the hawks, but it was more than that," says Walt Jendrzejewski. "There were a whole lot of causes, and they had a good healthy base to start from." Jendrzejewski does at least leave open the possibility that predator control had something to do with the irruption; some would deny it utterly. "Some ecologists accept the highly questionable theory that coyotes and similar predators do more good than harm because they help mother nature 'keep the balance,'" wrote *Oregonian* outdoor writer Don Holm not long ago. "It is a concept that appeals to many ill-informed but emotionally inclined citizens who have grown up with Walt Disney and television wildlife 'documentaries' . . . but anyone who comes from farming country knows that rodent populations control themselves, and predators have little or no effect on them." Citing "recent studies by experienced biologists," including one California investigation of three thousand coyote stomachs in which, says Holm, domestic animals and game "far outnumbered rabbits and rodents," he concludes: "In short, as a rodent controller, the coyote is a failure." And Holm's view, too, finds support from authorities, including Maurice Sheldon of Texas A & M's Agricultural Research and Extension Center in San

Angelo, Texas. "The coyote has been eliminated from the Edwards Plateau in Texas for over fifty years," points out Sheldon, in the December 1973 issue of *Bioscience* magazine, "but the area is not inundated by hordes of rodents." He terms the idea that coyotes help control rodent populations a "myth," and there are a number of other recent papers from such diverse locations as Alberta, California, and Nebraska, that agree with him.

Which view is right? Curiously enough, it turns out that *both* are: Predators assert both considerable control, and no control at all, over rodent numbers. To see how this seemingly impossible dichotomy can be resolved, though, we will have to delve a little deeper into the mechanics of predation as a natural phenomenon, and its role in the lives of both predators and prey.

We tend to think of predators as destructive creatures, and use the word "predatory" as a synonym for "destructive," "pillaging," or "rapacious"; but this is true only from a very limited perspective. Predation is invariably destructive to the animal being preyed upon—but, by the same token, it is invariably beneficial to the predator itself. And to the natural system of which both predator and prey are a part, it is neither destructive nor beneficial, but is merely a mechanism for keeping things going.

Predator-prey relationships almost always represent the end product of a process called *coevolution,* which may be defined as a simultaneous change in different species of animals or plants caused by their relationship to one another. It works this way: You begin with a pair of species that have developed a loose, accidental bond to each other, such as a plant that happens to exude a substance that some flying insect decides it likes to eat. Traveling from plant to plant in search of the substance to eat, the insect accidentally carries with it some of the pollen grains necessary for the plants to fertilize each other. Since even this accidental transport is more efficient than wind-carried pollen distribution, the plants that exude the nectar gain a reproductive advantage over the plants that don't, and begin to outnumber them. But the process doesn't stop there. Among the nectar-exuding plants, those that attract the insects most strongly will be able to outbreed their less fortunate brethren, and consequently attractive devices begin to appear: multi-colored petals, strong scents, copious amounts of nectar. At the same time, the flying insects develop senses that help them locate the attractants, such as color vision and extremely sensitive senses of smell. The plants develop pollen producers that are positioned where the insects must rub against them to get to the nectar; the insects develop hairy bodies that carry the pollen more efficiently and thus breed more of the plants they like to eat. The end result is, of course, bees and flowers, which have developed today to the point where they are totally dependent on one another for survival. If flowering plants were to die out, the bees would starve; if bees died out, the flowering plants would not be able to reproduce and would eventually die without issue. The two species have developed sets of characteristics together that have brought them into tighter and tighter dependence on one another. They have *coevolved*.

Coevolution can create dependencies that show up in unexpected places. English farmers discovered long ago, for instance, that if they were going to raise healthy beef cattle they also had to keep cats. The reasons for this rather curious relationship were pointed out by Charles Darwin as long ago as 1859 in *The Origin of Species*. English beef cattle are fattened primarily on red clover.

This clover has developed a close symbiotic relationship with bumblebees, which have become its exclusive pollinator: No other bee has a proboscis long enough to penetrate the clover flower's calyx and reach its nectar. Now, bumblebee colonies—unlike honeybee colonies—must be established anew each year, and there is a stage during the spring when the queen is the only adult around; while she is off foraging for food during this period, the young are defenseless, and become prey for mice, which apparently find them delectable. The cats keep the mouse population down, which keeps the bumblebee population up, which increases the clover's ability to pollinate, resulting in more clover and fatter cattle. The chain is rather long and roundabout, but every link is clear. In fact, when you add the fact that the cats are dependent on the farmers and the farmers are dependent on their cattle, the whole thing can be seen as a circle. Break it at any point—remove cats, or mice, or farmers, or cattle, or bumblebees, or clover—and the rest will very quickly get thrown out of balance.

A similar circular—or, more accurately, triangular—relationship has evolved in the old-growth forests of the Pacific Northwest, where it was explained to me in a conversation with old-growth specialist Dr. Glenn Juday at Oregon State University in Corvallis. The great old trees provide homes for a species of mouselike animals called red tree voles; the voles depend for their food, to a large degree, on a group of fungi known as the *micorrhiza*—popularly called truffles—which live entirely underground, attached to the roots of the great old trees. The trees depend on the micorrhiza to process certain soil nutrients, notably nitrogen, for them, and may die without their presence. The micorrhiza, in turn, depend on the voles to reproduce, for it is in the voles' digestive tracts that the fungus's spores get carried through the forest. And the voles depend on the trees for shelter. Break this triangular dependency at any one of its three nodes—voles, trees, or micorrhiza—and the remaining two will suffer severely.

Predator-prey relationships aren't often as complex as the cattle-clover-bumblebee-mouse-cat chain, or even the tree-vole-micorrhiza cycle. They tend to be two-way relationships, like that of the honeybee and the flower. Predators develop the tools necessary for pursuit, including a refined sense of smell, binocular vision, fangs and claws, and speed. Prey responds by developing tools of avoidance: a refined sense of hearing, wide-angle vision, agility, and a high birthrate to assure replacement of the animals the predators catch. It is this last-named adaptation, the high birthrate, that creates the mutual dependency between predators and prey. The predators depend on that high birthrate to keep them supplied with food; the prey depends on the hunting skill of the predators to keep that same high birthrate from suffocating them in their own kind. Kill off the prey population, and the predators will starve to death; kill off the predator population, and the prey will overpopulate its range, eat up all the food, and also starve to death.

The classic demonstration of what happens when a predator-prey symbiosis becomes upset took place during the early part of the twentieth century on the Kaibab plateau, perched spectacularly on the north lip of the Grand Canyon in the northwest corner of the state of Arizona. The Kaibab is a biotic island, eleven thousand square miles of pine forests and wildflower-laden glades surrounded on three sides by the mind-boggling gulf of the great canyon and on the fourth by one hundred miles of desert. In its natural state—the condition which prevailed up until about 1885—it is thought to have had the capacity to support between thirty and forty thousand deer. These deer would have been preyed upon principally by cougars, but also by wolves and possibly coyotes and bobcats. A fair number would also have gone into human stomachs, the Kaibab being a popular hunting ground for both the Navaho and Paiute Indians. The system appears to have been in a good dynamic balance, with the number of animals the predators (humans included) took from the population each year being approximately equaled by the number of new individuals added to that population by births during the same year. There was enough food for all.

In the mid-1880s, stockmen discovered the plateau and began moving cattle and sheep onto it. By 1900 there were perhaps fifty thousand domestic animals on the Kaibab. They competed for forage with the deer, and the deer herd dropped. It was down to under five thousand animals, and still shrinking, when Teddy Roosevelt stepped in. In 1906, a year before he created a waterfowl refuge out of Lower Klamath Lake, Roosevelt created a deer refuge out of the Kaibab, declaring the entire plateau a game preserve. The cattle and sheep were moved out, and government agents began trying energetically to rebuild the deer herd. In order to do this, they decided, the predators would have to go. There was already a healthy predator control program on the plateau, begun by the stockmen, but the wildlife people beefed it up. Predators virtually disappeared.

Now, deer—like most prey animals—have a high reproductive rate, with annual herd increases of 30 to 40 percent common and explosions of more than 100 percent in a single season not unheard of. In a balanced predator-prey population the predators would take those excess animals, but the Kaibab was no longer in balance. By 1918—barely twelve years after the program to rebuild the Kaibab herds had begun—the deer population was back up to around forty thousand animals and the forage, which had never caught up from the depletion caused by overgrazing by domestic livestock, was already scarce. Six years later, in 1924, the number of deer had reached one hundred thousand and the ecology of the plateau was in a shambles, its herbage all but gone, its shrubs dying, and its trees showing the typical browse-line formation of a severely depleted range, their branches chewed off in a uniform line precisely as high as the deer could reach. That winter was a severe one, limiting the browse still further. Sixty thousand deer starved to death.

The decrease continued over the next five years, with deaths—mostly from starvation—exceeding births, until the population had dropped, by 1931, to twenty thousand sickly animals on a decimated range. In that year the government stopped shooting predators and began shooting deer instead. By 1939 the herd was down to ten thousand animals and the predators had built back up to the point where they could once again begin asserting some control. Today, on the three-fourths of the plateau managed by the U.S. Forest Service, hunters and predators together keep the deer herd steady; on the remaining one-fourth, within the boundaries of Grand Canyon National Park, predators do the job alone. On both portions, the plant life has recovered. The system is back in balance, and a lesson—hopefully—has been learned.

If the Kaibab situation stood alone, as an isolated example, it might be possible to dismiss it as an aberration, a special case of predator-prey relationships that has no bearing on others. But it does not stand alone; wherever similar situations have existed, similar results have been seen. Near Lake Superior, a twelve hundred-acre fenced preserve owned by the University of Michigan experienced a similar boom-and-bust deer population cycle in the 1930s, shortly after deer were first introduced to the predator-free area within the fence. In Yellowstone National Park, at about the same time, predator control—practiced since the late nineteenth century—had to be halted because the elk herd had expanded to the point where it was practically eating the park whole. In Iowa, where the only large predators of note are humans, deer hunting was banned in 1936 with the state's deer population down to about six hundred animals. Fifteen years later, the herds had increased to at least eighteen thousand and almost as many deer were being killed by automobiles each year as had existed in the whole state when the ban was instituted.

In the world of insects, where predator-prey relationships tend to be highly specialized, the removal of predators can cause disasters similar to that of the Kaibab deer herd in a much shorter time. A team of San Diego scientists, for example, once undertook to study predator-prey relationships in an avocado orchard by systematically hand-removing all predatory and parasitic insects on one branch of an avocado tree. They did this for twelve weeks. At the end of that period the population of pest insects (caterpillars, mites, scale insects, and others) had risen so high that the alarmed scientists began hand-removing *them* to keep the branch from being defoliated. But on other trees in the same orchard—and even on other parts of the same tree in which the experiment was going on—no problem existed.

An excellent way to show the effects of upsetting the predator-prey balance in the insect world is to spray a crop with a pesticide to which the predators are susceptible and the prey is relatively immune. One example of this has already been mentioned—that of the cottony-cushion scale in the California citrus groves in the late 1940s (chapter 8), but it might not hurt to go over another. In *Biological Control by Natural Enemies,* biologist Paul DeBach relates his experiences with DDT and red scale, another citrus pest. He has tried spraying selected trees with DDT at monthly intervals throughout the growing season; the result is a dependable increase in the scale-insect population of anywhere from 36 to 1,250 times, while unsprayed trees in the same groves continue to show low to insignificant levels of the pest. "Most growers do not like this sort of test and have usually insisted that we stop before the trees were killed," reports DeBach, "but it is a wonderful way to raise red scale." Yellow scale responds in much the same way; it is, in fact, according to DeBach, "an easier way to raise yellow scale than it is in the insectary." These test results, along with their experiences with the cottony-cushion scale, have caused most citrus growers to virtually abandon the use of pesticides.

Mice, of course, are neither deer nor insects, and they have a different—and far less specialized—relationship to their predators. It takes a large and powerful predator to kill a deer, and insect predators are often highly specialized to penetrate the defense systems of only a single species of prey, but almost anything will eat a mouse. When I asked Steve Cross for a list of principal predators of the field mouse in Klamath Basin, he just looked at me and shook his head. "Principal predators," he muttered. "Mmmmm . . . . Well, raptors, coyotes, weasels, skunks—almost any predator, as a matter of fact. You can see herons mousing over there, walking around in the fields. I've watched shrikes feed on *Microtus,* and, heck, the *Microtus* is almost as large as the shrike." Gulls are also important ("It's *amazing* how big of a mouse a gull can swallow," says Walt Jendrzejewski), and Dayton Hyde reports that during the mouse irruption in the Klamath Basin his pet sandhill crane, Sandy, was instrumental in keeping the little creatures from overrunning his yard. "Voles were beginning to riddle the ditch banks with their tunnels, and to spoil the meadow," he wrote afterward, in his well-known book *Sandy.* "In the house lot, however, the tunnels ended abruptly, just a few feet inside the fence. Often as not, a fat black vole was left lying dead, a punctuation mark at the end of his run."

The voles have responded to all this predator pressure by developing a remarkably high birthrate. Females begin breeding at about six weeks of age, and during the remainder of their lives may continue to produce litters of up to nine young each as little as three weeks apart; William H. Burt, in his *Field Guide to Western Mammals,* tells of one captive female that had seventeen litters in a single year, an average of only 22.8 days between litters. It has been calculated that if the survival rate were 100 percent, a single pregnant female could become fifty thousand mature animals in one year's time. But—due in part to the great number and variety of predators, and in part to other factors such as disease and intraspecies fighting—this potential is ordinarily not even remotely approached. Though voles have a laboratory life span of as much as three years, they seldom live beyond nine months in the wild, and infant mortality is very high.

All those predators mean something else, too: They mean that the effect of fluctuations in the numbers of any single predator species will be next to nonexistent. When the prey species is one such as a deer or a cottony-cushion scale with only one or two major predators, the destruction of a single predator species will lead to a population explosion. When the prey animal has many predatory species using it, however, destruction of one of those species simply means more for the rest, and they move over to take up the slack. That is why predators are not generally thought to have much of a role in controlling those field mouse population cycles we discussed a few chapters ago. In fact, it works the other way: The field mouse fluctuations are almost certainly responsible for fluctuations in predator populations. "You'd have to say," is the way Cross puts it, "that the low point of the microtine cycle is the limiting factor for most predator populations."

There is considerable evidence to back him up, perhaps the most convincing piece of which is the population cycle of the Arctic fox, known from studies of fur trade returns compiled in the 1930s by Charles Elton. In Labrador, where the fox's chief prey is field mice and lemmings, the fur returns go up and down in a three-year pattern; farther west, on the Canadian mainland where the fox's chief prey is the snowshoe hare, fluctuations in the returns lengthen out to a nine-year pattern, which just happens to be the same length as the population cycle of the hare. If the predator cycles were controlling those of the prey, you would have to expect the length of the fox cycles to remain steady from place to place, with the vole/lemming cycle and the hare cycle both matched to it, and the fact that they do not remain steady is a powerful argument against giving the predator-prey interaction any credit at all for driving the basic small-mammal population cycles. Don Holm and his authorities are essentially correct.

And yet, it is equally wrong to assume that predators have no influence at all in small-mammal cycles: Hyde and his authorities are correct, too. The basic cycle must be driven by something else, but there is ample evidence to suggest that the *amplitude* of that cycle—how high or low it gets—is at least partially controlled by predators.

Some of the best of this evidence comes from places where small mammals with high reproductive rates have been introduced into environments with few or no predators. Australia, as we have seen, has had extensive troubles with introduced populations of European rabbits; Puerto Rico has had similar problems with rats. On the San Juan Islands in Washington's Puget Sound, introduced rabbits have multiplied so greatly in the absence of predators that their numerous tunnels have begun to weaken the islands' structure, and large pieces of island have begun falling off into the sea.

Other evidence comes from places where the predators, once present, have been exterminated. The Scottish Border Hills, for example, had no history of large-scale rodent plagues before the nineteenth century. In the early years of that century, a bounty system on all predators (including birds of prey), plus a flourishing fur trade, virtually wiped out the predator population; exhibition of polecat skins at the Dumfries Fair, to pick one indicator out of many, declined from six hundred in 1831 to twelve in 1866, and none at all in 1870. During the next twenty-five years—in 1872, and again in 1892—the Border Hills experienced two of the largest vole plagues in history.

It may help if I interject about here something that Ed O'Neill told me, right after he had finished pulling my leg about the sunspots. At least, I think he was pulling my leg. I'm not sure. Anyway, what he said was, "Actually, we haven't wondered so much about what *causes* irruptions, but we'd sure as hell like to know what *prevents* 'em. 'Cause that would really be a good management tool." Looked at that way, everything suddenly clicks into place. We know the cause of irruptions: It is simply the extraordinary reproductive powers of the field mouse during the increase phase of its cycle. The real question is not where those powers come from, but why they don't lead to an irruption every three years. And the answer to that is really pretty clear, too.

Something must be killing off the excess mice.

What kills a mouse? Lots of things. Weather can do it, if it is hot enough to cause heat exhaustion or cold enough to cause hypothermia. Starvation and thirst can do it. Psychological stress can kill a mouse—as it can a human— and so can the presence of toxic chemicals in the environment, either intentional (poison) or accidental (pollution). But by far the biggest factors are two we have already talked about: disease and predation.

The evidence we have looked at so far has all come from irruptions that have taken place in the absence of predators. I'd like to close this chapter, therefore, with two quotes from studies made of *nonirruptive* populations. The first does not actually have much to do with mice; it comes from the report of an eight-year investigation of jackrabbits in Utah by Frederick H. Wagner and Charles L. Stoddart. "The data indicate," write Wagner and Stoddart (*Journal of Wildlife Management,* April 1972), "that 69 percent of the observed variation in rabbit numbers is associated with variation in the coyote:rabbit ratio. Accordingly, we postulate that coyote predation played an important role in the jackrabbit population trends from 1962 to 1970, hastening, if not primarily causing, the decline from 1963 to 1967 by its impact, and largely, or in part, permitting the increase in rabbits in 1968–70 by its relaxation." The second quote comes from the study of Arctic lemmings by F. A. Pitelka and his colleagues, which we quoted briefly earlier in this chapter; and because lemming and field mouse cycles occur simultaneously wherever the two animals are found together it *does* have to do with mice. Most of Pitelka's conclusions, as we saw, dealt with the effect of snow depth on lemming reproduction. But he also studied predation, especially predation by weasels and Arctic foxes, and it is the results of that part of the study that we want to look at now. Writing in the Winter 1974 issue of *Arctic and Alpine Research,* Pitelka said: "We believe that predation, in particular predation by resident carnivores lacking sufficient alternative prey, may exert severe pressure on a declining lemming population. *This predation may be important in determining the amplitude of the cycle by forcing and maintaining the depth of the low. This crucial predation need not occur at the point of maximum prey density."* (Emphasis added.)

Predation *does* control animal numbers, even if it doesn't drive population cycles. Predators *are* important.

12

LET US PAUSE briefly to consider where we are.

We have begun with a natural population of field mice, or voles—*Microtus montanus*—occupying the moist grasslands of the Klamath Basin and cycling their populations up and down every three to four years in the so-called microtine cycle, for what cause scientists are still not certain.

We have drained two ninety thousand-acre lakes and turned large portions of their former beds into irrigated cropland, thus expanding the voles' habitat. They have responded by taking the first opportunity—the 1920 population peak—to explode into their new living quarters. There is too much land for the old predator population to patrol, and the explosion gets out of hand, forcing us to take over some of the predators' role. We do this by poisoning the ditch banks every year. The vole population settles back fairly dependably within the limits of its normal cycle.

Beginning about 1945, we launched an all-out war on the coyote population. Statewide, this has succeeded in dropping the numbers of these important predators by about 60 percent, or roughly 600 animals in the Klamath Basin. Since each coyote disposes of about 1,200 mice per year (the coyote's average annual food intake in pounds times percentage of the coyote's diet known to be field mice times number of field mice per pound), this decline of 600 coyotes means a potential increase of about 720,000 mice—if something else doesn't eat them.

Fortunately, something else does. Coyotes are not the only creatures that eat mice, and some of the increase is being picked up by other predators such as bobcats, skunks, weasels, and badgers. However, the same predator control program responsible for eliminating the coyote is also going after a lot of the coyote's competition: A typical predator-control score sheet for one month in Klamath County at the time of the mouse irruption logs the unnatural end of twenty-six bobcats, five badgers, and one skunk, along with the nine coyotes that represent the program's primary aim. Because of this, these other predators are at best just holding their own, and the slack left by the decrease in coyotes will have to be taken up somewhere else—by avian predators such as gulls, herons, and raptors (hawks and owls), and by human poisoning programs.

The poisoning programs are fairly predictable. They will stay at an established level until there is a demonstrated need to increase, and since this demonstration can only consist of unacceptably high levels of damage, by the time the increase in poisoning comes the irruption will already have taken place. So that leaves the birds. What about the birds?

There is only one thing to do. We will have to go find Ed O'Neill, over there at the Tule Lake Wildlife Refuge, where they keep track of Klamath Basin bird populations, and ask him.

Ed was working on something else when I came, unannounced, through his office door. He sat at a desk mounded with paperwork, the sleeves rolled up on his Fish and Wildlife Service khaki shirt. I was obviously going to be an interruption he didn't really want. However, there was no one else who would have the information I needed, so I braced him for it anyway. What, I wanted to know, could he tell me about raptor population trends during the past thirty years?

He gave me a rather glum look. "Raptors?" he said. "Down."

I must have looked disappointed at his brevity, because he immediately relented. Somewhere beneath that gruff exterior there is a heart of gold. "Well," he corrected himself, throwing down his pencil, "the resident raptors really haven't fluctuated much recently. We've had a stable population since I've been here, and I came here in 1961. But you'd have to say the raptor population dropped a lot before I came. When I first saw the place, back in forty-nine, there were probably four inches of droppings over on the peninsula, in the petroglyph section of Lava Beds National Monument and up on the cliffs there above it, where the owls nest. None of that's there now. And back then folks were saying, 'That's nothing, you should have seen it *when*.' So—yeah. The raptor population is *way* down. But the figures we have are mighty sketchy. You see, now we look at all the birds, but back in those days nobody cared much about predators. It was all waterfowl."

What, I wondered, did "sketchy" mean? Was there any data at all?

"Here," he offered, "I'll show you." He leaped to his feet, suddenly full of energy, and flew into a filing cabinet in one corner of his office, coming up with a thickly stuffed manila folder of data. "Some of my own," he explained, flipping the folder open in front of me. "That

way I can't be accused of criticizing someone else's work."
I looked at the top sheet of the bundle in the folder. It was
a three-month bird count report for January–April 1976,
listing some 270 raptors with the count broken down into
species—so many short-eared owls, so many golden eagles,
so many red-tailed hawks, so many marsh hawks, and so
forth. What, I wondered, was sketchy about this? Ed
laughed.

"Don't lean too heavily on those figures," he warned
me. "What we do is, we turn in weekly reports, and one
week you might count twenty owls, another week maybe
twenty-one, and you infer from that that there's maybe
fifty out there, so that's what you put down here." He
tapped the report with a blunt forefinger.

"Well," I said, "that's certainly better than nothing. I
mean, it would give an indication, wouldn't it, of population trends, even if the figures themselves aren't too
reliable? If the estimates went way down, wouldn't that
be a pretty good indicator that the population was going
down, too?"

He nodded. "Yes, I suppose it would."

"Well, then, do you have any sheets like this for anything earlier?"

"How early?"

"Oh, say, back about nineteen-fifty."

"Nineteen-fifty." He inclined his head musingly. "That's
mighty sketchy, my friend, mighty sketchy. . . ." He rifled
through a few more drawers, pulling out manila folders,
glancing at them, and stuffing them back in. Then he
suddenly slammed the cabinet shut and started out the
door. "Let's try the basement," he called over his shoulder.
I corralled my notebook and hurried down the hall after
him.

The basement was dark and musty and full of odd shapes

and corners. One of those corners held a little room about eight feet square whose walls—floor to ceiling, left to right —were solid filing cabinets. "Nineteen-fifty," said Ed. "Hmmm. . . ." He chose a cabinet, pulled it open, yanked out a folder, and suddenly the figures were before me.

> January–April, 1950: 473 raptors sighted
> January–April, 1953: 317 raptors sighted
> January–April, 1955: 160 raptors sighted
> January–April, 1956: 74 raptors sighted

The data might be sketchy, but there could be no mistake about what it meant. During the period 1950–56, the raptor population of the Klamath Basin was decreasing. Drastically decreasing, in fact, if the estimates are to be believed: A decline from 473 birds to 74 birds is a drop of almost 85 percent. At a time when the populations of birds of prey should have been increasing to take over the role of the vanishing coyotes, they were going downhill even faster than the coyotes.

What caused this decline? O'Neill refused to speculate, but I think I know. I think the clue lies in an incident that took place four years later, in the summer of 1961, right there at the Tule Lake National Wildlife Refuge. We have already told the story of that incident, back in chapter 8—of how, during the course of that long summer, hundreds of birds were found dead on the refuge grounds, victims of what autopsies proved to be pesticide poisoning. Pesticides, especially DDT, cause reproductive failure in birds long before they reach levels high enough to cause actual deaths. The mechanism through which this operates is well known: It consists of an interference with the way in which the birds metabolize calcium, causing them to lay very thin-shelled eggs. The thin-shelled eggs crack very easily. The warm weight of the parent bird, meant to incubate the egg to life, crushes it to death instead.

There were few raptors in the Klamath Basin, it now seems almost certain, because almost none were being born.

And now, at last, we can put the whole picture together. We cannot know what causes *all* irruptions, but we can be almost certain of what caused the great irruption in the Klamath country in 1957–58. We don't have to look far for the cause: A mirror will do. We did it to ourselves.

The coyote populations were depleted—killed purposefully by man to protect his flocks.

The bird of prey populations were depleted—killed accidentally by man as a by-product of pesticide spraying to protect his crops.

The flea and louse populations were depleted—killed accidentally by man in the same way he killed the birds.

A typical periodic high was due in the microtine cycle: There were no predators, there was no disease, and the winter was mild and warm. The Klamath Basin had thousands of the healthiest, safest, most fecund mice you ever saw. And the population of those mice grew, and grew, and grew. . . .

And maybe there was another factor operating, too. I got a hint of what it might be from one old farmer who refused to let me use his name, but who was clearly upset by the whole thing. This man runs a five thousand-acre cattle ranch on the upper Lost River, in the Poe Valley area, one of the two worst hit regions of the Klamath Basin during the mouse irruption. He has been there for over twenty years. "In the beginning," he told me, of that Year of the Mice, "we had a lot of poison scattered pretty indiscriminately around, and it got a lot of the predators. And without the predators to help us, why, it just got away. It was just preposterous, the lack of supervision we had on the broadcast of poison."

The deadly grain meant for the mice was getting their enemies, too. As in the case of Danysz's Virus in France, the cure was not only worse than the disease—it was contributing to it. The foul-ups we had created in the bird and coyote and flea populations were being compounded by our attempts to rectify them, and we were being overrun by the products of our own ignorance.

13

In baiting a mousetrap with cheese, always leave room for
the mouse.

Saki (H. H. Munro), *The Square Egg,* 1924

THE LONG-AWAITED mass poisoning program finally
got under way on February 27, 1958. Seven days later,
on March 5, the California portion of the program was
abruptly halted and a meeting of wildlife and agriculture
officials from all three Klamath Basin counties was hastily
assembled at the headquarters building on the Tule Lake
Refuge. Five hundred geese, most of them the beautiful
and popular snow geese, had been found dead on the
refuge, and heads were going to roll.

The poisoning was a many-pronged effort. The ditch
banks were being baited by irrigation district personnel—
Klamath Irrigation District in Oregon, Tule Lake Irrigation
District in California. The road edges were being handled
by the county roads departments in California; Harold
Schieferstein, the Klamath County weed and rodent
supervisor, was in charge of the road-edge baiting program
in Oregon. Farmers in both states had been issued poison
bait at or below cost, and they were placing it in their
own fields. Zinc phosphide bait was being used in Oregon
and in Siskiyou County, while Modoc County was going

with 1080. Recommendations for both baits were similar: Place the treated grains by hand in piles twenty feet apart, at a rate not to exceed ten pounds per acre, and cover the piles with a little straw so that game birds will not be attracted to them. But the recommendations did not have the force of law, and somebody out there wasn't following them.

The goose deaths were quickly traced to an eighty-eight acre patch of fields in the Panhandle region, an arm of the old Tule Lake bed that stretches eastward and northward from the vicinity of the northeast corner of Lava Beds National Monument. Here, as in many other places, the hand-baiting recommendation had been ignored, and the poison—six hundred pounds of zinc phosphide and another 280 pounds of 1080—had been scattered instead from a low-flying airplane. The Southeast Sump area of the Tule Lake Wildlife Refuge is immediately adjacent to the Panhandle, separated from it only by an elevated roadway that leads into the national monument, and the spring migration of geese was swelling toward its height. The thousands of incoming geese had found their food gone, eaten by the mice—and there was all that tempting grain scattered all over the surface of the ground next door. It was supposed to be colored bright red to warn the birds away, but most of it wasn't. They came, and they ate. And they died.

The coalition of agriculture and wildlife officials at the refuge headquarters meeting reached agreement quickly on a new, stronger set of recommendations: These were issued within days of the discovery of the dead geese. They emphasized the need for caution, reminding farmers once again of the requirement for color-coding, stipulating that baiting should be done only in fields where cover was adequate to keep flying birds from spotting the poisoned grain at all, and suggesting that it might be well to wait a few weeks before doing any more baiting, to let the peak of the northern goose migration pass. In the meantime, the refuge staff would go to a supplementary feeding program, spreading grain within the refuge to attract the geese inward and keep them from venturing into the lethal fields. It was a good program, but it was also, as the more cynical among the officials had suspected it would be, a flat failure. It is probable that most farmers followed the recommendations, but a large enough number ignored them so that their effect was pretty well nullified. Little attention was paid to the request to postpone the baiting until after the goose migration, though this had been a standard practice in other years. ("Regular baiting was always done after the birds had left and before they returned," Walt Jendrzejewski told me, "but when they exploded and were eating everything up, it was bait whenever you could, I guess.") Aerial application continued to be a major tool of the poisoners; others used fertilizer spreaders or "grain drills"—mechanical planters whose normal job is to sow the seeds of life, not death. The recommended dosage of ten pounds of poisoned grain per acre treated was routinely doubled and sometimes tripled. All in all, an estimated 300,000 pounds of death—nearly 150 tons—rained onto the fields of the Klamath Basin in a period of time probably not exceeding four weeks in length.

The geese continued to die. A pilot operating a baiting plane on the lower Williamson River neglected to turn off his spreader as he made a turn over the riverbank, and a hunter in the area a few days later found thirty dead Canada geese; the poison application rate in this case was put down as the recommended ten pounds per acre, but the official report of the incident notes rather laconically that "rates considerably above ten pounds were probable." Sixty-four dead geese were found on the Lost River on March 15; forty-one on Spring Lake on March 18; and seventy-five on Alkalai Lake on March 20. Thirty-three were found together in a single field near Merrill on March 22, eleven in a ditch at the south end and the remaining twenty-two nearby. One field near Henley apparently accounted for numerous deaths: Though only two geese were actually found in the field, many others were discovered scattered between it and the nearest body of water, their corpses forming a straight line along the logical flight path. When the figures were added up, confirmed goose poisonings for that one-month period in the Klamath Basin reached 3,676, a number that one biologist estimated "probably does not include more than two-thirds of the actual loss." The geese had literally been decimated: One out of every ten flying north through the Klamath that spring had died.

What about the mice themselves—the target of all those tons of poison? Local officials hailed the program as a success, with 25 to 50 percent kills noted "at a single poisoning," and the *Herald and News* and the *Oregonian* ran articles duly reporting this opinion, along with pictures of fields full of dead mice. There was, however, conflicting evidence as to how accurate those pictures were as a measure of success. Some of the more skeptical biologists pointed out that a large dieback could be expected at this time of the year anyway, from purely natural causes, and that back in mid-February—well before the mass poisoning program was officially off the ground—this natural dieback had already been estimated at up to 50 percent, or approximately the amount of the highest reported poison kill. It was also noted that, given the high reproductive potential of the mice, a 50 percent kill could be entirely nullified by the birthrate in less than a month, if that birthrate were not being offset by a high *continuing* death rate. In the short term, the program was unquestionably a success; in the long term, it is debatable whether it had any effect at all.

The birds were better. They had rebounded amazingly from their population nadir two years previously; the Tule Lake Refuge's spring raptor tally for 1958 was almost double that of 1956, and the local Audubon Society's Christmas bird count showed an even more striking upward leap, from a low of 160 raptors (1955) to a high of 438 (1957), an increase of 275 percent and the highest number of hawks and owls seen in the basin since 1951. Gull numbers soared even more spectacularly, from a low of 40 individuals in the 1955 Audubon count to a high of 1,520 in 1957—a whopping 3,800 percent population climb. Most of this increase was undoubtedly due not to reproduction but to flocking: The freedom of the air gives avian predators a high degree of mobility, and they will come into an area where prey pickings are good from as far as several hundred miles away. How they learn about areas of abnormally good hunting is one of the great mysteries of ornithology, but it appears obvious that they have some means of communicating the information to each other. Consider, for example, the strange case of the disappearing gulls. Those 1,520 individuals counted at Christmas suddenly evaporated in early January, and all through the rest of that month and the first part of February their numbers were so few that nobody bothered to tally them. Then, suddenly, on February 10, they returned —with reinforcements. Great flocks of them, thousands strong, swept in over the mountains and circled into the basin as if it were a gigantic bull's-eye. Six thousand seven hundred fifty were counted along one eighteen-mile stretch of road near Malin—an average of 375 a mile. Other counts, in the Poe Valley and on roads leading south out of Klamath Falls, were similar. Somehow, clearly, the word had been passed that the Klamath country was a heck of a good place for a gull to be.

No count is available for the numbers of mice destroyed by all these birds, but it must have been upwards of several hundred thousand. Birds have high metabolic rates, and they require a lot of food. One researcher made a casual investigation of forty-five consecutive fence posts in the Poe Valley during the irruption, and found field mouse remains at the bottom of forty-two of them. Fence posts are known to be favored perching places for hawks and owls. Another researcher found a nest of great horned owls, which he managed to watch for seven straight weeks during the nesting season; during those seven weeks the parent birds and their four young did away with twenty-three hundred mice, a figure which indicates that mouse deaths by raptors alone must have been in the neighborhood of one hundred fifty to two hundred thousand. And the gulls were even better. "The gulls were a lot of help," says Walt Jendrzejewski. "You'd turn up a bale of hay and there'd be three or four mice under it, and the gulls were right there." As the fields were flooded for irrigation near Tule Lake the gulls gathered, watching the edge of the advancing water and pouncing on mice as they were forced out of their burrows. At Fort Klamath, a lingering snow-pack developed circular bare patches eight or ten inches in diameter, each patch marking a place where the warmth of a mouse nest had melted the snow; the gulls discovered these bare spots and homed in on them, tearing up the ground to expose the nests and eating every mouse in sight, dead or alive. "Generally half a dozen or more gulls were involved in killing a mouse," wrote John Craighead, observing in the basin for the U.S. Fish and Wildlife Service. "The normal procedure was for a gull to fly or run at a fleeing mouse, pick it up in its bill, then immediately release it. The same gull or another one would then repeat the performance. A mouse might be caught five or six times before it ceased struggling. When the mouse was dead or helpless, the last gull to grasp it would fly off attempting to swallow it in flight. Other gulls would attempt to steal the mouse. Eventually one gull among the flock would swallow the mouse intact." Craighead makes no estimates, but the number of mice who met their end in this rather gruesome manner must have been absolutely astronomical.

Still, it wasn't the birds that finally turned the tide, any more than it was the poison. It was something much smaller than either a gull or a piece of poisoned grain—a tiny creature called *Pasteurella tularensis,* better known as the tularemia microbe. "It was the tularemia that finally did 'em in," says Paul Hatchett, in a voice that brooks no argument. Was it really that effective? Hatchett chuckles. "Oh, yeah," he says. "You see, when the tularemia broke and we got word of it, the people who didn't have any went to the area where it was and picked up two or three tularemia mice, and so it spread fast." There were still few fleas around to aid that spread, but they were no longer needed. There is no mystery about how the disease was transmitted in the absence of vectors: The mouth lesions and enlarged cervical nodes of most of the victims point clearly to the means of transfer. It was cannibalism. The mice, having eaten up everything else in sight, had turned to eating each other.

Epilogue

IN LATE MARCH 1958, the Klamath Union High School basketball team won the state title, displacing the mice from the front page of the *Herald and News*. They never regained it. The plague was essentially over; mouse levels had dropped to below one hundred per acre everywhere, and were still declining. Slowly, farming in the Klamath Basin settled back to normal. Between 10 and 15 percent of the alfalfa and alsike clover stands were rated a total loss and had to be ploughed up and replanted. The plows and harrows moved slowly back and forth through the fields; the remaining mice, their burrows torn up, fled to the surface, where they were pounced on and swallowed by the clouds of waiting gulls—as many as fifteen hundred to a single flock—that followed the tractors on their routes. The farmers decided to like gulls. The normal crop rotation patterns were upset for many areas, but when the planting was done at the end of April the amount of land devoted to potatoes—18,500 acres—was about the same as in an average year. *Herald and News* editor Bill Jenkins reported that the mice in his yard were "healthy enough to eat the lawn, roots and all, cut down most of the flowers, dig up bulbs and just play hop in general." But the basin was recovering.

When the farmers went into the fields for first haying in mid-June, it was discovered that certain fields in the Tule Lake area which for some inexplicable reason had been relatively mouse-free during the height of the plague had high mouse populations now. There was a brief flurry of activity. The newspapers trumpeted an alarm about a "second wave" consisting of hordes of hungry mice "moving inland from the basin perimeter at the rate of one mile per week," and the poisoners got our their zinc phosphide and their 1080 and went to work. But the number of infested fields remained small, and the publicity bubble collapsed. As it turned out, this "second wave" was just the forerunner of what would come to be seen as a regular pattern, with abnormally high numbers of mice showing up somewhere in the basin almost every year from that time on. "I suppose the worst was at Fort Klamath about five years after the main irruption," recalls Paul Hatchett, rubbing his chin. "There were two or three thousand acres infested then. You know that curve on the highway about a quarter of a mile north of Fort Klamath, going toward Crater Lake? Well they were in the meadows on one side of that curve, and then it rained and the meadows got soggy, so they started to cross from one side to the other and the cars hit 'em. Got so bad the highway department had to go out with a road grader and scrape 'em off." Hatchett also recalls other small-scale infestations—near Henley, and in the Poe Valley, and on the grounds of the Klamath Experiment Station itself. Meanwhile, over the mountains to the east, in Malheur and Harney Counties, they were plagued with rabbits: They appeared suddenly in the fall of 1958, many thousand strong, chomping their way across the fields near Ontario, on the state's eastern border—the domain of coyote master-killer Pud Long—like a living wave. "There are thousands of them," said Mel Smith, district agent of the U.S. Fish and Wildlife Service. "It's impossible to estimate the population. They're all over. You see them on the roads. You see them in the fields." Asked by a reporter for the *Oregonian* to estimate their numbers, Leeds Bailey, county agent for Malheur County, simply shrugged his shoulders. "Take any number," he said. "It's hard to estimate." But the rabbits, too, died back, and nothing since has even remotely approached the scale of the great Klamath Mouse War. Gradually, the image of a land being eaten by hordes of hungry rodents has faded into memory.

The Klamath Basin today shows little sign of having once been the scene of the greatest mouse irruption in the history of North America. The fields are neat and groomed; machinery and buildings look well cared for, and the farms have a prosperous appearance. There is some mouse damage every year, but they have learned to live with it and to minimize it. Where once irrigation was done with a single winter flooding of the fields, leaving the mice free to breed in perfect habitat for the rest of the year, the farmers now flood the fields several times each season; each time gives them a chance to bait more mice. Where once the combines left six- to eight-foot swaths uncut at the edge of each field, today's machines can cut to within inches of the fence posts, leaving the mice with only a fraction of the cover they once enjoyed and making them much more vulnerable to predators. "Out in the middle of a field, you know, a mouse hasn't got a chance," says Paul Hatchett with obvious satisfaction. Predators themselves have increased somewhat, as predator control has been relaxed a little, following the national aversion to mass coyote-killing that arose in the early 1970s. Spraying has also decreased, and the most dangerous varieties of spray—the chlorinated hydrocarbons—have virtually disappeared, leaving the mice once again subject to diseases carried by vector insects. One county in the basin—Modoc —has taken the bull by the horns, or the mouse by the ears, and formed a Mouse Abatement District—which I suppose abbreviates as MAD, though I've not seen the term used. "They've had it for the last half-dozen years or so," Ed O'Neill told me. "They assess each landowner something like a cent an acre, and then use the money to purchase and mix bait. And they're convinced they're controlling irruptions that way, and there hasn't been one in Modoc County since the program started. But there hasn't really been one in Siskiyou or Klamath County, either, so it's kind of hard to tell. I don't really know." He leaves unspoken the fervent wish of every Klamath Basin resident: *I don't really want to find out.* One record mouse irruption was quite enough. There is no need to put the basin on the map in quite that manner for a second time.

Few residents of the Klamath country appear to have changed their ways to any great degree as a result of the Great Mouse War. There are, however, some notable exceptions. One of these exceptions is Dayton Hyde. "The mouse irruption was really an eye-opener for me," he says today. "I saw very suddenly that every county agent in the world was wrong, and that they just had no idea what they were trying to do." He decided to try ranching as if nature were an ally rather than an enemy to be subjugated. Most of his eight thousand-acre ranch beneath Yamsi Mountain was drained marshland; he let 25 percent of it revert to marsh and, because these new wetlands attracted insect-eating birds that kept his crop pests down, he saw his cattle production go up by 54 percent. Experiments with captive coyotes convinced him that the rodents each one ate during the hundred-day growing season would have destroyed one hundred dollars worth of forage crops and that therefore, with lamb prices standing at forty dollars a head, each coyote could kill two lambs during that one hundred-day period and society would still owe him twenty dollars. Predator control was henceforth banned from his ranch. "We've also quit trying to poison the short-tailed ground squirrels," he relates. "As long as we were poisoning them, we had healthy populations. I finally realized that what was happening was that the poison would knock the population down to the point where the predators would all leave, and then without that control on it it would just spring back up again. We quit poisoning and went to natural predator control, and now we've got a fairly stable population of ground squirrels at probably twenty-five percent of what we had when we were poisoning." Currently, Hyde is serving on the national board of directors of Defenders of Wildlife, the anti-predator control organization, converting two thousand acres of his ranch into a private wildlife refuge, and harboring a dream of reintroducing wolves into the Klamath country. "Wouldn't it be wonderful," he muses, "to have wolf howls soaring over these hills again?"

Others do not share this vision. They still shoot predators in Klamath County. They still spray their fields, and though the sprays they use are far less persistent and dangerous, there is a strong undercurrent of support for a return to the good old days of DDT, endrin, toxaphene, and the other chlorinated hydrocarbons. Hyde, whose books on ranching with nature have made him a prominent target, has had to suffer a considerable amount of harassment. "I couldn't get a permit to have a coyote from Fish and Game for a long time," he says, grinning wryly. "The sheepmen weren't *about* to let me have one, because I said in my permit application I was looking for the good side of the coyote. Everyone else is looking at the bad side, I figured someone ought to look at the good side for a change." In the early 1970s he went to Washington, D.C. to testify against predator control at a series of congressional hearings, citing his own experiences in the great mouse irruption and since. He left home vice-chairman of the Oregon Cattlemen's Association; he returned "drummed out of the organization. I got thirty-nine hate calls in one night." A group of sheepmen attacked him in the hall outside the Senate hearings room, and the resulting scuffle had to be broken up by Capitol guards. Change, even in mouse-eaten Klamath County, is not going to come easily.

Hyde remains philosophical about the whole thing. "I will never make coyote lovers out of my neighbors who are coyote haters," he wrote, a few years ago. "[But] I think that we must take a clear look at predators, and while we may add up what they do against us, we must also recognize what they do for us." The same clear look, he emphasizes in private conversations, must be taken at pesticides and at poisoning programs in general. While there may be some advantages to some of these things, these advantages must always be weighed against the harmful effects. And their use, while it need not be entirely eradicated, must always be limited by the tight and very real boundaries of those harmful effects. Nature *is* in balance: It is subject, as a whole, to that "delicate equilibrium" that Charles Elton noted long ago in his mice, where "quite a small variation . . . would cause surprisingly large changes." The great Klamath mouse irruption should have

172

taught us what those small variations can do. Where is the lesson that has been learned?

We sit in Hyde's cabin near the Williamson River on a cold February Sunday. Half of the living room is shut off behind floor-to-ceiling barricades of chicken wire, and behind the chicken wire Hyde's collection of tropical songbirds chirps and twitters and darts about with flashes of brilliant color. At his feet lounges a pet wolf; beyond, over his shoulder, the lake he is creating on his private refuge spreads a wide gray sheet of water across what used to be a drained and channeled pasture. "I feel that all life is going somewhere, from single-celled animals on up," he muses, swirling the tea about in his big mug. "It's directed. And this business of man having dominion—I can't *stand* that." How long, I wonder, will it be until all men share Hyde's vision? How many Mouse Wars will it take, how many chewed-up fields, how many destroyed ditch banks? How many mice can dance on the head of a pin?

There is a balance of nature, and it gives us three choices. We can ignore that balance, make changes arbitrarily on one side or the other, and watch while the balance is restored by uncontrolled changes on the other side; this, the Klamath mouse irruption should have taught us, is clearly unwise. We can make no changes at all, on either side; this would take us back a hundred thousand years, back to before the agricultural revolution, and leave us gathering nuts and berries and bringing down game with our teeth, and that is also clearly unwise. Or we can make changes, carefully, on *both* sides, utilizing natural processes instead of fighting them, preserving the balance *and* shaping the world. This is the most difficult choice, far harder than either of the others. But it is also the only one worth looking at.

The Carson Factor

It all boils down once again to the Carson Factor—the awareness that we are dealing with life. John Muir expressed it well a century ago: "When you try to pick out anything by itself," he wrote, in what is probably the best one-sentence description of ecology ever devised, "you find it hitched to everything else in the Universe." Rachel Carson herself put the same thought in slightly different words. Life, she told us in *Silent Spring*, is a fabric—"a fabric on the one hand delicate and destructible, on the other miraculously tough and resilient, and capable of striking back in unexpected ways." It is our great misfortune as a people to have been caught in an age when we have achieved the capacity to destroy that fabric—and continue to lack the ability to perceive, at any one time, more than a few inches of a single thread.

A Selected Bibliography

BOOKS

Andrewartha, H. G. *Introduction to the Study of Animal Populations*. Chicago: The University of Chicago Press, 1961.

Aristotle. *Historia Animalium*. English translation by A. L. Peck. Cambridge, Mass.: Harvard University Press, 1970.

Boyle, John C. *50 Years on the Klamath*. Klamath Falls, Oregon: Klamath County Museum, 1976.

B. R. *The Famous Hystory of Herodotus*. New York: AMS Press, Inc., 1967.

Brewer, E. Cobham. *Dictionary of Phrase and Fable*. Philadelphia: J. B. Lippincott Company, 1894.

Burt, William H. *Field Guide to Western Mammals*. Boston: Houghton Mifflin Company, 1964.

Carson, Rachel. *Silent Spring*. Boston: Houghton Mifflin Company, 1962.

DeBach, Paul. *Biological Control by Natural Enemies*. London: Cambridge University Press, 1974.

Dicken, Samuel N. *Oregon Geography: The People, the Place, and the Time.* 4th ed. Eugene, Or.: University of Oregon Cooperative Bookstores, 1965.

Elton, Charles. *Voles, Mice and Lemmings: Problems in Population Dynamics.* Oxford: The Clarendon Press, 1942.

Emmel, Thomas C. *An Introduction to Ecology and Population Biology.* New York: W. W. Norton & Company, Inc., 1973.

Graham, Frank Jr. *Since Silent Spring.* Boston: Houghton Mifflin Company, 1970.

Hazen, William E., ed. *Readings in Population and Community Ecology.* Philadelphia: W. B. Saunders Co., 1970.

Hyde, Dayton O. *Sandy.* New York: The Dial Press, 1968.

Kendeigh, S. Charles. *Animal Ecology.* Englewood Cliffs, N.J.: Prentice-Hall, Inc., 1961.

___. *Klamath Basin.* Salem, Oregon: Oregon State Water Resources Board, 1971.

___. *Klamath County, Oregon, Resource Atlas 1973.* Corvallis, Oregon: Oregon State University Federal Cooperative Extension Service, 1973.

Leydet, Francois. *The Coyote: Defiant Songdog of the West.* San Francisco: Chronicle Books, Inc., 1977.

Maser, Chris, and Storm, Robert M. *A Key to Microtinae of the Pacific Northwest.* Corvallis, Oregon: O.S.U. Book Stores, Inc., 1970.

Milne, Lorus, J., and Milne, Margery. *The Balance of Nature.* New York: Alfred A. Knopf, Inc., 1960.

Murie, Adolph. *Ecology of the Coyote in the Yellowstone.* Washington, D.C.: United States Department of the Interior, National Park Service, 1940.

___. *The Oregon Meadow Mouse Irruption of 1957-1958.* Corvallis,

Oregon: Federal Cooperative Extension Service, Oregon State College, 1958.

Pimentel, David. *Ecological Effects of Pesticides on Non-Target Species.* Washington, D.C.: Executive Office of the President, Office of Science and Technology, 1971.

Ryden, Hope. *God's Dog.* New York: Coward, McCann & Geohegan, Inc., 1975.

Sisemore, Linsey, ed. *History of Klamath County.* Klamath Falls, Oregon: (n.p.), 1941.

Sperry, Charles C. *Food Habits of the Coyote.* Washington, D.C.: United States Department of the Interior, Fish and Wildlife Service, 1941.

Van Wormer, Joe. *The World of the Coyote.* Philadelphia: J. B. Lippincott Company, 1964.

Vaughan, Terry A. *Mammalogy.* Philadelphia: W. B. Saunders Company, 1978.

PERIODICALS

Agriculture Bulletin (Oregon State Department of Agriculture). March 1946, and March 1947.

Biennial Report of the Oregon State Department of Agriculture. 1940-57.

Connoly, Guy E., Timm, Robert M., Howard, Walker E., and Longhurst, William M. "Sheep Killing Behavior of Captive Coyotes." *Journal of Wildlife Management,* July 1976.

Guthery, Fred S., and Beasom, Samuel L. "Responses of Game and Nongame Wildlife to Predator Control in South Texas." *Journal of Range Management,* November 1977.

Hyde, Dayton O. "Man's Best Friend—The Coyote." *Defenders of Wildlife,* June 1974.

The Carson Factor

Klamath Falls Herald and News, 1957–58.

Knowlton, Frederick F. "Preliminary Interpretations of Coyote Population Mechanics with Some Management Implications." *Journal of Wildlife Management,* April 1972.

Krebs, Charles J. "Demographic Changes in Fluctuating Populations of *Microtus californicus.*" *Ecological Monographs,* Summer 1966.

Krebs, Charles J., Gaines, Michael S., Keller, Barry L., Myers, Judith H., and Tamarin, Robert H. "Population Cycles in Small Rodents." *Science,* January 5, 1973.

Maize, Kennedy P. "The Great Kern County Mouse War." *Audubon,* November 1977.

MacLean, S. F. Jr., Fitzgerald, B. M., and Pitelka, F. A. "Population Cycles in Arctic Lemmings: Winter Reproduction and Predation by Weasels." *Arctic and Alpine Research,* Winter 1974.

Negus, Norman C., and Berger, Patricia J. "Experimental Triggering of Reproduction in a Natural Population of *Microtus montanus.*" *Science,* June 10, 1977.

Oregonian (Portland, Oregon). 1912–71.

Ryden, Hope. "Shearing the Public, and the Clamor to Poison the Coyote." *The New York Times,* July 29, 1978.

Shelton, Maurice. "Some Myths Concerning the Coyote as a Livestock Predator." *Bioscience,* December 1973.

Wagner, Frederick H., and Stoddart, L. Charles. "Influence of Coyote Predation on Black-Tailed Jackrabbit Populations in Utah." *Journal of Wildlife Management,* April 1972.

Wynne-Edwards, V. C. "Population Control in Animals." *Scientific American,* August 1964.

Notes

PREFACE

vi. "Paul Erlich": Erlich, *The Population Bomb* (New York: Ballantine Books, Inc., 1968).

vi. "Rachel Carson": *Silent Spring* (Boston: Houghton Mifflin Company, 1962).

vi. "Annual Report of the El Paso Electric Company": quoted in *The Chat* (Rogue Valley Audubon Society), November 1978, p. 2.

viii. "V. C. Wynne-Edwards": as expressed in articles such as his "Population Control in Animals," *Scientific American,* August 1964.

viii. "What Rachel Carson called a 'state of adjustment'": Carson, *Silent Spring,* p. 17.

ix. "Charles Elton . . . flinging a crowbar": in Charles Elton, *Voles, Mice, and Lemmings: Problems in Population Dynamics* (Oxford: The Clarendon Press, 1942), p. 105.

ix. "Carson writes": Carson, *Silent Spring,* p. 261.

xii. "Murie's extremely hard-to-get book on coyotes": Adolph Murie, *Ecology of the Coyote in the Yellowstone, Fauna of the National Parks of the United States.* Bulletin No. 4 (Washington, D.C.: United States Department of the Interior, National Park Service, 1940).

The Carson Factor

PROLOGUE

1. "a bit of a fable": The following account is compiled from several sources, notably *Encyclopaedia Brittanica,* 15th ed. (Chicago: Encyclopaedia Brittanica, Inc., 1974); *Webster's Biographical Dictionary* (Springfield, Mass: G. & C. Merriam Co., 1974); E. Cobham Brewer, *Dictionary of Phrase and Fable* (Philadelphia: J. B. Lippincott Company, 1894); *Larousse Encyclopedia of Mythology* (London: Hamlyn Publishing Group, Ltd., 1968); and the *New Century Cyclopedia of Names* (New York: Appleton-Century Crofts, Inc., 1936).

4. "Robert Southey": Southey, "The Legend of Bishop Hatto," in Lieder, Lovett, and Root's *British Poetry and Prose.* (Boston: Houghton Mifflin Company, 1938), vol. 2, p. 343.

5. "Thomas Coryat": quoted in Emily Morison Beck, ed., *Bartlett's Familiar Quotations,* 14th ed. (Boston: Little, Brown and Company, 1968), p. 532n.

6. "E. Cobham Brewer": See note to p. 1.

1

7. "On Sunday, January 20": Unless otherwise noted, all accounts of events during the years 1957–58 in the Klamath Basin are based on articles in the *Klamath Falls Herald and News* during those two years. These notes will not refer to the *Herald and News* further except in the case of direct quotes or statistics: In such cases, the abbreviation *KFH&N* will be used.

2

12. "The poet Joaquin Miller": quoted in Harry Hansen, ed., *California: A Guide to the Golden State* (New York: Hastings House, 1967), p. 433.

13. "As of 1978, the official population": Information in this and the next four sentences was provided by the Klamath County Chamber of Commerce.

13. "This is quite possibly the finest agricultural land": Except as noted directly below, the remaining information in this paragraph comes from *Statistical Abstract of the United States,* 98th ed. (Washington, D.C.: U.S. Bureau of the Census, 1977), and the *Klamath County Resource Atlas* (Corvallis, Oregon: Oregon State University Extension Service, 1973).

14. "Even less acreage is planted to clover": The data and quotation in this sentence are from the *KFH&N,* November 5, 1957.

3

15. "an old wildlife magazine in my den": Dayton Hyde, "Man's Best Friend—The Coyote," *Defenders of Wildlife,* June 1974.

17. "one publication, put together by scientists": *The Oregon Meadow Mouse Irruption of 1957-1958* (Corvallis, Oregon: Federal Cooperative Extension Service, Oregon State College, 1958), hereinafter cited as *OMMI.*

19. "from the Old Norse *vollmus*": A Scottish Presbyterian minister turned naturalist named John Fleming is usually credited with the first use of the word *vole* in English: He picked it up among the inhabitants of his parish in the Orkney Islands, where it was a dialectical holdover from the days when the Orkneys were ruled by the Vikings. See *Oxford English Dictionary* (Oxford, England: The Clarendon Press, 1933).

20. "Dr. Robert M. Storm": In Chris Maser and Robert M. Storm, *A Key to Microtinae of the Pacific Northwest* (Corvallis, Oregon: Oregon State University Bookstores, Inc., 1970), p. 17.

4

26. "the Klamath country looked very different": See, e.g., Linsey Sisemore, ed., *History of Klamath County, Oregon* (Klamath Falls, Oregon, 1941. No publisher given, but probably Western Historical Publishing Company); and *The Klamath Country,* (Southern Pacific Railway Company, 1908), a promotional pamphlet.

27. The Roosevelt quotation on this page, and the information about the Reclamation Act following it, are from Theodore Roosevelt, *An Autobiography* (New York: Charles Scribner's Sons, 1926), p. 396.

27. "There was already a fair amount of land under irrigation": This and the following information about Klamath Basin irrigation up through the Modoc Project comes from Sisemore, *History of Klamath County,* pp. 103-115; and John C. Boyle, *50 Years on the Klamath* (Klamath Falls, Oregon: Klamath County Museum, 1976).

28. "The single comic-opera meeting": This anecdote is retold from Sisemore, *History of Klamath County*, p. 111.

5

36. The Jendrzejewski quote is from *KFH&N*, May 11, 1958.
38. "mice are very destructive": *KFH&N*, July 7, 1957.
39. "excellent mouse control": *KFH&N*, July 7, 1957.
39. "Some fields that had shown extremely high potential yields": *OMMI*, p. 5.
41. "In numerous places large potatoes": *OMMI*, p. 6.
41. "oat/strychnine bait as 'very effective'": *KFH&N*, September 22, 1957.
42. "In the Langell Valley": *OMMI*, p. 7.
43. "Bill Jenkins reported": *KFH&N*, April 9, 1958.
43. "At Fort Klamath": *OMMI*, p. 6.
43. "All over the basin": *OMMI*, p. 8.
44. "scurrying over the surface": *KFH&N*, March 7, 1958.
44. "'frequently . . . can be observed crossing the highways'": *KFH&N*, February 6, 1958.
44. "'riddled' ditch banks and drain banks with 'tunnels'": *OMMI*, p. 7.
44. "One official talley near Tule Lake": Both tallies in this sentence are found in *OMMI*, p. 21.

6

47. "In southern Idaho": *OMMI*, p. 19.
47. "on Sauvies Island": This information, and the Tom Davis quotation, come from the Portland *Oregonian*, December 18, 1957. Hereinafter cited as *OREG*.
48. "In the northern Idaho town of Moscow": *KFH&N*, January 23, 1958.
48. "Mice on the Gale Day farm": *KFH&N*, December 13, 1957.
48. "Editor Bill Jenkins waxed philosophical": The waxing took place in *KFH&N*, January 21, 1958.
49. "rancher W. H. Steiwer": *OREG*, November 27, 1957.
50. "one staff writer": Don Holm, "Adaptability Helps Coyote Survive Attempts at Extermination," *OREG*, May 24, 1971.
51. "Some ecologists go so far": e.g., Thomas C. Emmel, *An Introduction to Ecology and Population Biology* (New York: W. W. Norton & Company, Inc., 1973), p. 39.

51. "the 'coyote pest'": All three of these terms appeared in *Oregonian* articles (October 20, 1912; January 16, 1944; and February 11, 1947).

52. "gleefully wire captured coyotes' mouths shut": These and other incidents are told in Francois Leydet's excellent book *The Coyote: Defiant Songdog of the West* (San Francisco: Chronicle Books, 1977).

52. "Robert E. 'Pud' Long of Odessa": *OREG,* January 26, 1947.

53. "just . . . for the lust of killing": *OREG,* October 20, 1912.

53. "Mark Twain": Mark Twain, *Roughing It* (New York: Harper & Brothers Publishers, 1913), p. 32.

54. "Charles C. Sperry": Charles C. Sperry, *Food Habits of the Coyote* (Washington, D.C.: United States Department of the Interior, Fish and Wildlife Service, 1941).

55. "Adolph Murie in Yellowstone": See note to p. xii.

55. "Hope Ryden": Ryden, *God's Dog* (New York: Coward, McCann & Geohegan, Inc., 1975).

55. "Frederick Wagner and Charles Stoddart": Wagner and Stoddart, "Influence of Coyote Predation on Blacktail Jackrabbit Populations in Utah," *Journal of Wildlife Management* 36:2, 1972.

55. "The report of the Cain commission": quoted in Leydet, *Songdog of the West,* p. 31. Leydet also has a good summary of the Leopold Commission findings.

55. "One group of researchers": Guy E. Connally, Robert M. Timm, Walker E. Howard and William M. Longhurst, "Sheep Killing Behavior of Captive Coyotes," *Journal of Wildlife Management* 40:3, 1976.

56. "two areas in south Texas": Fred S. Guthery and Samuel L. Beasom, "Responses of Game and Nongame Wildlife to Predator Control in South Texas," *Journal of Range Management* 30:6, 1977.

57. "as many as 75 percent of reported 'kills'": Leydet, *Songdog of the West,* p. 123.

58. "a coyote hunting technique": See Joe Van Wormer, *The World of the Coyote* (New York: J. B. Lippincott Company, 1964).

58. "the number of sheep raised in this nation": These figures are assembled from two sources, *Colliers Encyclopedia* (New York:

Macmillan Educational Corporation, 1978), Vol. 20, p. 657; and *KFH&N*, January 6, 1957.

59. "Set out poisoned animal carcasses": Leydet, *Songdog of the West,* pp. 126–128.
59. "Hunt him by air": *OREG*, March 17, 1947.
59. "One pair of researchers": Leydet, *Songdog of the West,* p. 152.
59. "'Don Coyote'": *OREG*, April 26, 1944.
60. "AIR WAR": The headlines in this and the next two sentences appeared, in order, in *OREG*, March 17, 1947; February 14, 1947; and February 23, 1947.
60. "'Eradication of the coyote'": *OREG*, February 14, 1947.
61. "'Extermination, rather than mere control": *OREG*, March 25, 1945.
61. "the landing at Plymouth Rock": William Bradford, *History of Plymouth Plantation,* ed. William T. Davis (New York: Barnes & Noble, Inc., 1959), p. 96.
61. "reaching Oregon in 1848": *Oregon Agricultural Bulletin,* March 1946.
61. "The first modern bounty law": *OREG*, October 20, 1912.
62. "In 1914": Leydet, *Songdog of the West,* p. 105.
63. "in the state of Oregon alone": *OREG*, February 18, 1947.
63. "a million and a half coyotes dead": *OREG*, January 16, 1944, gives the figure as 1,465,575.
63. "figures published by the old Predator and Rodent Control Branch": These figures are a bit hard to obtain (Leydet devotes a whole chapter to what he calls "The Numbers Game"—the difficulty of obtaining reliable statistics). But see the Oregon State Department of Agriculture's *Biennial Reports* for the years 1940 through 1957.
64. "dropping approximately 60 percent": *OREG*, February 14, 1947.
64. "the 'coyote-getter'": See Leydet, *Songdog of the West,* pp. 108, 127.
65. "a substance called sodium monofluoracetate": *Ibid.,* pp. 109–111.
65. "Originated in South Dakota": *OREG*, March 25, 1945.
65. "a mechanic named Floyd Capp": *OREG*, April 26, 1944.
66. "aim for the tip of the tail": *OREG*, March 4, 1945.

66. "Oregon Game Commission . . . three crews of its own": *OREG*, March 4, 1945.
66. "PARC agents . . . into the act": *OREG*, March 17, 1947.
67. "PARC kill reports for Oregon": See note to p. 63.
67. "equations used by population biologists": See, e.g., H. G. Andrewartha, *Introduction to the Study of Animal Populations* (Chicago: University of Chicago Press, 1961), pp. 31-34.
67. "'natural' population density": Frederick F. Knowlton, "Preliminary Interpretations of Coyote Population Mechanics with some Management Implications," *Journal of Wildlife Management* 36:3, 1972.
68. "Deer herds had increased": *KFH&N*, February 15, 1957 and June 20, 1958.
68. "Edwin J. Casebeer": *KFH&N*, January 2, 1957.
68. "The Weyerhaeuser Company": *KFH&N*, March 26, 1958.

7

69. "a special report to the U.S. Public Health Service": quoted in *OREG*, November 17, 1957 and November 21, 1957.
71. "Deputy Agricultural Commissioner Bill Huse": *KFH&N*, November 22, 1957.
72. "a special meeting of . . . health officials": reported in *OREG*, November 17, 1957.
73. "'antimouse disease'": *OREG*, November 21, 1957.
74. "The newspapers announced it": *OREG* and *KFH&N*, November 21, 1957.
74. "One official went so far as to suggest": *OMMI*, p. 46.
75. "summed up Fritz Bell": *KFH&N*, December 11, 1957.
75. "remarked Vertrees": *KFH&N*, December 11, 1957.
76. "Prinz agreed . . . Prinz admitted": *OREG*, December 11, 1957.
76. "the governor pointed out": *OREG*, December 18, 1957.
76. "Fish and Wildlife . . . announced somewhat grumpily": *KFH&N*, December 20, 1957.
77. "Robert J. Seward, announced": *OREG*, December 19, 1957.
77. "editorial cartoon": *OREG*, December 1, 1957.
77. "Hatfield issued": *OREG*, December 19, 1957.
78. "Seward muttered darkly": *KFH&N*, December 19, 1957.
78. "most important plant pests": *Oregon Agriculture Bulletin*

No. 197 (Salem: Oregon State Department of Agriculture, March 1958).

78. "the governor announced": *OREG,* December 25, 1957.

8

81. *"Newsweek* picked up the story": *Newsweek,* January 6, 1958.
82. "a massive program . . . to fight the screw-worm fly": This story is told in numerous places. See, e.g., Paul DeBach, *Biological Control by Natural Enemies* (London: Cambridge University Press, 1974), p. 23.
83. "the Great Australian Rabbit Invasion": Like the account of the screw-worm fly, this story is widely available. My principal source for both the Australian events and the French and English follow-ups was Lorus J. and Margery Milne, *The Balance of Nature* (New York: Alfred A. Knopf, Inc., 1960), chapter 13.
86. "Charles Elton . . . ground-breaking studies": Elton, *Voles, Mice, and Lemmings,* p. 194.
87. "the strange case of the Danysz' Virus": Danysz appears to be one of those figures that biology would just as soon forget. But see Elton, *Voles, Mice, and Lemmings,* pp. 30–47.
89. "the report of the pathologists from the U.S. Public Health Communicable Diseases Center": *OMMI,* pp. 43–53.
90. "The *Oregonian* thought it knew": *OREG,* November 23, 1957.
91. "the Russian researcher M. R. Netsengevitch": *Biological Abstracts* 35:6, 1960.
93. "The recommendations": The list on this and the next few pages was prepared by correlating chemical and brand-name recommendations from the weekly "County Agents Report" column in *KFH&N* for January 1957 through December 1958 with information about uses, chemical structure, and effects of those chemicals from several sources, notably the "County Agents Reports" themselves; *The McGraw-Hill Encyclopedia of Science and Technology* (New York: McGraw-Hill Book Company, 1977); *Grzimek's Animal Life Encyclopedia* (New York: Van Nostrand Reinhold Company, 1975);

and interviews with chemist Rodney A. Badger at Southern Oregon State College, Ashland, Oregon.

96. "CMU . . . would kill small fish": David Pimentel, *Ecological Effects of Pesticides on Non-Target Species* (Washington, D.C.: Executive Office of the President, Office of Science and Technology, 1971), p. 114.

97. "ads placed by . . . the Veliscol Chemical Company": *KFH&N*, April 3, 1958.

98. "the citrus farmers of California's Owens Valley": DeBach, *Biological Control*, pp. 2-10.

98. "the worst in twenty years": *KFH&N*, July 7, 1957.

99. "damning evidence of pesticide overuse": Carson, *Silent Spring*, pp. 49-50.

99. "the *Oregonian's* . . . pesticide theory": See note to p. 90.

9

101. "L. A. Peterson of Mist": *OREG*, January 9, 1958.

101. "Mrs. Jean White": *OREG*, December 16, 1957.

101. "the Albany . . . *Democrat-Herald*": *KFH&N*, September 9, 1958.

102. "A. Noni Mouse": *KFH&N*, December 18, 1957.

103. "'All-Out War'": *KFH&N*, December 29, 1957.

103. "one U.S. Fish and Wildlife Service biologist": Donald A. Spencer in *OMMI*, p. 22.

107. "According to Henderson": *KFH&N*, February 9, 1958.

107. "Bill Decker . . . reported": *KFH&N*, January 9, 1958.

107. "The Public Health Service": *KFH&N*, December 15, 1957.

107. "the conferees stated": *OREG*, February 13, 1958.

108. "worried the county agent's office": *KFH&N*, February 27, 1958.

108. "fretted the *Herald and News*": *KFH&N*, February 23, 1958.

10

109. "The reproduction of mice": Aristotle, *Historia Animalium*, A. L. Peck (Cambridge, Mass: Harvard University Press, 1970), Book 6, p. 349.

109. "The Old Testament": I Samuel 6:5.

109. "Diodorus Siculus, Strabo, and Pliny": cf. Elton, *Voles, Mice, and Lemmings*, p. 4.

109. "Herodotus": B. R., *The Famous Hystory of Herodotus* (London, 1584, Facsimile ed. New York: AMS Press, Inc., 1967), p. 220.

109. "pregnant merely by licking salt": Aristotle, *Historia*, p. 349. The statue of Apollo is mentioned in Elton, *Voles, Mice, and Lemmings*, p. 5.

110. "John Stow's *Chronicles*": quoted in Elton, *Voles, Mice, and Lemmings*, p. 7. The material following the quotation, describing medieval methods of combatting mouse plagues, is likewise from Elton (chapters 1 and 2).

111. "The German naturalist I. H. Blasius": *Ibid.*, p. 53.

111. "The Frenchman Charles Gerard": *Ibid.*, p. 1.

111. "The British Board of Agriculture . . . one typical report": *Ibid.*, pp. 8, 126–140.

113. "The Kern County outbreak": *Ibid.*, pp. 108 ff, and Kennedy P. Maize, "The Great Kern County Mouse War," *Audubon*, November 1977, pp. 159ff.

114. "no lessons were learned": See chapter 13.

116. "some of Christian's work": e.g., J. J. Christian, "The Adreno-Pituitary System and Population Cycles in Mammals," *Journal of Mammalogy*, May 1950; and J. J. Christian and D. E. Davis, "Endocrines, Behavior, and Population," *Science*, December 18, 1964.

11

122. "Terry A. Vaughan's *Mammalogy*": Terry A. Vaughan, *Mammalogy* (Philadelphia: W. B. Saunders Company, 1978), p. 321.

122. "S. E. Piper's now-classic study": quoted in Elton, *Voles, Mice, and Lemmings*, p. 108.

123. "The same point . . . by Charles Elton": *Ibid.*, p. 25.

127. "one biologist": Ed O'Niell of the Tule Lake National Wildlife Refuge, quoted extensively later in this book.

127. "Donald A. Spencer": *KFH&N*, January 25, 1958.

127. "Alice Gay": *KFH&N*, January 8, 1958.

129. "I looked up a number of sources": e.g., David Lack, *The Natural Regulation of Animal Numbers* (Oxford: The Clarendon Press, 1954), p. 212; and A. MacFayden, *Animal Ecology* (London: Sir Isaac Pitman & Sons, Ltd., 1957), p. 208. Neither

of these sources, incidentally, thought much of the sunspot theory.

129. "Charles Elton showed": Elton, *Voles, Mice, and Lemmings,* p. 160.

129. "S. Charles Kendeigh": S. Charles Kendeigh, *Animal Ecology* (Englewood Cliffs, N. J.: Prentice-Hall, Inc., 1961), pp. 243-244.

130. "A. H. R. Goldie": Elton, *Voles, Mice, and Lemmings,* p. 190.

130. "F. A. Pitelka": S. F. MacLean, Jr., B. M. Fitzgerald, and F. A. Pitelka, "Population Cycles in Arctic Lemmings: Winter Reproduction and Predation by Weasels," *Arctic and Alpine Research* 6:1, 1974.

130. "Elton himself discovered": Elton, *Voles, Mice, and Lemmings,* p. 181.

131. "Negus managed to isolate a substance": Norman C. Negus and Patricia J. Berger, "Experimental Triggering of Reproduction in a Natural Population of Microtus Montanus," *Science,* June 10, 1977.

131. "temperatures in the Klamath Basin": as reported in *KFH&N,* November 1956 through February 1957.

132. "a very delicate equilibrium": Elton, *Voles, Mice, and Lemmings,* p. 204.

136. "Charles Elton tells us": See note to p. 86.

137. "Without coyotes to hold them in check": Hope Ryden, "Shearing the Public, and the Clamor to Poison the Coyote" *The New York Times,* July 29, 1978.

137. "Carson says it": Carson, *Silent Spring,* p. 219.

137. "*Oregonian* outdoor writer Don Holm": Don Holm, "Adaptability Helps Coyote Survive Attempts at Extermination," *OREG,* May 24, 1971.

138. "points out Sheldon": Maurice Sheldon, "Some Myths Concerning the Coyote as a Livestock Predator" *Bioscience,* December 1973.

141. "the Kaibab Plateau": This story is probably told in more ecology textbooks and general-interest environmental accounts than any other predator/prey balance story. My account is based largely on that in Milne and Milne, pp. 78-82.

143. "fenced preserve owned by the University of Michigan": The

George Reserve. See Milne and Milne, *Balance of Nature,* pp. 71–74.

144. "In Yellowstone National Park": Murie, *Ecology of the Coyote,* pp. 10, 11–16, 57.

144. "In Iowa, where only": Milne and Milne, *Balance of Nature,* p. 85.

144. "a team of San Diego scientists": *Ibid.,* p. 20.

145. "DeBach relates his experiences": DeBach, *Biological Control,* pp. 2–10.

146. "'voles were beginning to riddle the ditch banks'": Dayton O. Hyde, *Sandy* (New York: The Dial Press, Inc., 1968), p. 158.

146. "William H. Burt": William H. Burt, *Field Guide to Western Mammals* (Boston: Houghton Mifflin Company, 1964), p. 192.

146. "it has been calculated": *OREG,* November 21, 1957.

146. "seldom live beyond nine months": Elton, *Voles, Mice, and Lemmings,* pp. 110–112, 173–176, 202.

147. "population cycle of the Artic fox . . . in Laborador . . . on the Canadian mainland": *Ibid.,* chapter 11.

148. "Australia, as we have seen . . . rabbits": See note to p. 83. Australia has had mouse troubles, too, but they are essentially unrelated to those of the Klamath. Australian "mouse plagues" have been caused by house mice (*Mus musculus*) rather than by field mice. Like the rabbit, it has been a case of an imported species with a high birth rate coming into an area with few predators and a virtually unlimited food supply. The results have occasionally been astounding: During a plague of this nature in 1917, for instance, one farmer poisoned his front porch, retired for the night, and shoveled 28,000 carcasses off the porch the next morning. See Elton, *Voles, Mice, and Lemmings,* pp. 100–101; also "Mice Threat 'Down Under'" *Science Newsletter,* September 26, 1959.

148. "Puerto Rico has had similar . . . rats": Milne and Milne, *Balance of Nature,* pp. 42–54.

148. "On the San Juan Islands": *Ibid.,* chapter 13.

148. "The Scottish Border Hills": Elton, *Voles, Mice, and Lemmings,* pp. 126–140.

149. "Wagner and Stoddart": See third note to p. 55.

149. "F. A. Pitelka and his colleagues": See third note to p. 130.

12

151. "take over some of the predators' role": For a cogent comparison of poisoning programs with predation, see Elton, *Voles, Mice, and Lemmings*, p. 59.

152. "each coyote . . . 1,200 mice per year": Each adult coyote eats about 2.5 pounds of food per day (Hyde, "Man's Best Friend"). About 20 percent of this, conservatively estimated, will be field mice (pp. 54–55 and notes thereto); since mice come about 6.5 to the pound (Burt, *Field Guide*), we get $(2.5 \times 365) \times 20\% \times 6.5 = 1,186.25$.

152. "a typical predator-control score sheet": *KFH&N*, May 5, 1958.

155. "the mechanism through which this operates": Frank Graham, Jr., *Since Silent Spring* (Boston: Houghton Mifflin Company, 1970), pp. 127–130.

13

159. "When baiting a mousetrap": Saki (H. H. Munro), *The Short Stories of Saki: Complete with an Introduction by Christopher Morley* (New York: The Viking Press, Inc., 1934), p. 623.

159. "Five-hundred geese . . . had been found dead": *KFH&N*, March 6, 1958.

160. "Recommendations for both baits": *KFH&N*, February 27, 1958; *OMMI*, p. 11.

160. "The goose deaths": *KFH&N*, March 6, 1958.

160. "The coalition . . . reached agreement": *KFH&N*, March 7, 1958.

161. "recommended dosage of ten pounds": *KFH&N*, March 25, 1958.

161. "an estimated 300,000 pounds": *OMMI*, p. 12.

161. "The geese continued to die": *OMMI*, pp. 31–34.

161. "one biologist estimated": The biologist was James Mohr of the Oregon State Gam Commission. See *OMMI*, p. 33.

162. "'at a single poisoning'": *KFH&N*, April 27, 1958.

162. "Some of the more skeptical biologists": *KFH&N*, March 5, 1958.

162. "a 50 percent kill could be entirely nullified": *KFH&N*, December 11, 1957.

163. "the Tule Lake Refuge's spring raptor tally": Obtained from Ed O'Niell. See text, pp. 152–155.

163. "the Audubon Society's Christmas bird count": *KFH&N*, January 5, 1958.

163. "the strange case of the disappearing gulls": *OMMI*, pp. 37–38.

163. "Six-thousand-seven-hundred-fifty were counted . . . other counts": *OMMI*, p. 38.

164. "Forty-five consecutive fence posts": *OMMI*, p. 38.

164. "a nest of great horned owls": *KFH&N*, May 12, 1958.

164. "the gulls were even better": Except for the Jendrzejewski quote, which came from a personal conversation, the information in the rest of this paragraph comes from *OMMI*, pp. 39–40.

165. "mouth lesions . . . cannibalism": *KFH&N*, December 15, 1957; *OMMI*, p. 47. Dayton Hyde also noted in a personal conversation that the mice seemed to be "a seething mass of squealing bodies eating each other . . . they just turned cannibalistic."

167. "clouds of waiting gulls": *OMMI*, pp. 38–39.

167. "editor Bill Jenkins reported": *KFH&N*, April 9, 1958.

168. "the newspapers trumpeted an alarm": *KFH&N*, July 2, 1958 and July 9, 1958.

168. "Malheur and Harney Counties . . . plagued with rabbits": This information, including the two quotes, comes from *OREG*, October 23, 1958.

170. "25 percent of it revert to marsh": personal conversation with Dayton Hyde.

171. "Experiments with captive coyotes": Hyde, "Man's Best Friend."

172. "he wrote, a few years ago.": in the article that set me off on all this, by Hyde, "Man's Best Friend."

172. "Charles Elton noted long ago": See note to p. 132.

174. "John Muir expressed it well": John Muir, quoted in David Brower, ed., *Gentle Wilderness: The Sierra Nevada.* (San Francisco: Sierra Club Books, 1967), p. 146.

174. "Rachel Carson": Carson, *Silent Spring*, p. 261.

Index

1/94/3

1/11/1

I. Title.

QL737
.C22A8

Ashworth, William.
The Carson factor.

F37119

10.95